TM

References for the Rest of Us

COMPUTER BOOK SERIES FROM IDG

Are you baffled and bewildered by programming? Does it seem like an impenetrable puzzle? Do you find that traditional manuals are overloaded with technical terms you don't understand? Do you want to know how to get your PC to do what you want? Then the *...For Dummies* programming book series from IDG is for you.

...For Dummies programming books are written for frustrated computer users who know they really aren't dumb but find that programming, with its unique vocabulary and logic, makes them feel helpless. *...For Dummies* programming books use a humorous approach and a down-to-earth style to diffuse fears and build confidence. Lighthearted but not lightweight, these books are a perfect survival guide for first-time programmers or anyone learning a new environment.

"Simple, clear, and concise. Just what I needed."
—Steve P., Greenville, SC

"Finally, someone made learning to program easy and entertaining. Thanks!"
—Diane W., Chicago, IL

"When I saw this book I decided to give programming one last try. And I'm glad I did!"
—Paul G., St. Louis, MO

Millions of satisfied readers have made *...For Dummies* books the #1 introductory-level computer book series and have written asking for more. So if you're looking for a fun and easy way to learn about computers, look to *...For Dummies* books to give you a helping hand.

ISDN FOR DUMMIES™

ISDN FOR DUMMIES™

David Angell

IDG Books Worldwide, Inc.
An International Data Group Company

Foster City, CA • Chicago, IL • Indianapolis, IN • Braintree, MA • Dallas, TX

ISDN For Dummies

Published by
IDG Books Worldwide, Inc.
An International Data Group Company
919 East Hillsdale Boulevard, Suite 400
Foster City, CA 94404

Library of Congress Catalog Card No.: 95-79566

ISBN 1-56884-331-3

Printed in the United States of America

First Printing, June, 1995

10 9 8 7 6 5 4 3 2 1

Distributed in the United States by IDG Books Worldwide, Inc.

For More Information...

For general information on IDG Books in the U.S., including information on discounts and premiums, contact IDG Books at 800-434-3422.

For information on where to purchase IDG's books outside the U.S., contact Christina Turner at 415-655-3022.

For information on translations, contact Marc Jeffrey Mikulich, Foreign Rights Manager, at IDG Books Worldwide; fax number: 415-655-3295.

For sales inquiries and special prices for bulk quantities, contact Tony Real at 800-434-3422 or 415-655-3048.

For information on using IDG's books in the classroom and ordering examination copies, contact Jim Kelly at 800-434-2086.

The ...*For Dummies* book series is distributed in Canada by Macmillan of Canada, a Division of Canada Publishing Corporation; by Computer and Technical Books in Miami, Florida, for South America and the Caribbean; by Longman Singapore in Singapore, Malaysia, Thailand, and Korea; by Toppan Co. Ltd. in Japan; by Asia Computerworld in Hong Kong; by Woodslane Pty. Ltd. in Australia and New Zealand; and by Transword Publishers Ltd. in the U.K. and Europe.

Welcome to the world of IDG Books Worldwide.

IDG Books Worldwide, Inc. is a subsidiary of International Data Group, the world's largest publisher of computer-related information and the leading global provider of information services on information technology. IDG was founded more than 25 years ago and now employs more than 7,500 people worldwide. IDG publishes more than 235 computer publications in 67 countries (see listing below). More than fifty million people read one or more IDG publications each month.

Launched in 1990, IDG Books Worldwide is today the #1 publisher of best-selling computer books in the United States. We are proud to have received 3 awards from the Computer Press Association in recognition of editorial excellence, and our best-selling *...For Dummies*™ series has more than 18 million copies in print with translations in 24 languages. IDG Books, through a recent joint venture with IDG's Hi-Tech Beijing, became the first U.S. publisher to publish a computer book in the People's Republic of China. In record time, IDG Books has become the first choice for millions of readers around the world who want to learn how to better manage their businesses.

Our mission is simple: Every IDG book is designed to bring extra value and skill-building instructions to the reader. Our books are written by experts who understand and care about our readers. The knowledge base of our editorial staff comes from years of experience in publishing, education, and journalism — experience which we use to produce books for the '90s. In short, we care about books, so we attract the best people. We devote special attention to details such as audience, interior design, use of icons, and illustrations. And because we use an efficient process of authoring, editing, and desktop publishing our books electronically, we can spend more time ensuring superior content and spend less time on the technicalities of making books.

You can count on our commitment to deliver high-quality books at competitive prices on topics consumers want to read about. At IDG, we value quality, and we have been delivering quality for more than 25 years. You'll find no better book on a subject than an IDG book.

John Kilcullen
President and CEO
IDG Books Worldwide, Inc.

IDG Books Worldwide, Inc. is a subsidiary of International Data Group, the world's largest publisher of computer-related information and the leading global provider of information services on information technology. International Data Group publishes over 235 computer publications in 67 countries. More than fifty million people read one or more International Data Group publications each month. The officers are Patrick J. McGovern, Founder and Board Chairman; Kelly Conlin, President; Jim Casella, Chief Operating Officer. International Data Group's publications include: **ARGENTINA'S** Computerworld Argentina, Infoworld Argentina; **AUSTRALIA'S** Computerworld Australia, Computer Living, Australian PC World, Australian Macworld, Network World, Mobile Business Australia, Publish!, Reseller, IDG Sources; **AUSTRIA'S** Computerwelt Oesterreich, PC Test; **BELGIUM'S** Data News (CW); **BOLIVIA'S** Computerworld; **BRAZIL'S** Computerworld, Connections, Game Power, Mundo Unix, PC World, Publish, Super Game; **BULGARIA'S** Computerworld Bulgaria, PC & Mac World Bulgaria, Network World Bulgaria; **CANADA'S** CIO Canada, Computerworld Canada, InfoCanada, Network World Canada, Reseller; **CHILE'S** Computerworld Chile, Informatica; **COLOMBIA'S** Computerworld Colombia, PC World; **COSTA RICA'S** PC World; **CZECH REPUBLIC'S** Computerworld, Elektronika, PC World; **DENMARK'S** Communications World, Computerworld Danmark, Computerworld Focus, Macintosh Produktkatalog, Macworld Danmark, PC World Danmark, PC Produktguide, Tech World, Windows World; **ECUADOR'S** PC World Ecuador; **EGYPT'S** Computerworld (CW) Middle East, PC World Middle East; **FINLAND'S** MikroPC, Tietoviikko, Tietoverkko; **FRANCE'S** Distributique, GOLDEN MAC, InfoPC, Le Guide du Monde Informatique, Le Monde Informatique, Telecoms & Reseaux; **GERMANY'S** Computerwoche, Computerwoche Focus, Computerwoche Extra, Electronic Entertainment, Gamepro, Information Management, Macwelt, Netzwelt, PC Welt, Publish, Publish; **GREECE'S** Publish & Macworld; **HONG KONG'S** Computerworld Hong Kong, PC World Hong Kong; **HUNGARY'S** Computerworld SZT, PC World; **INDIA'S** Computers & Communications; **INDONESIA'S** Info Komputer; **IRELAND'S** ComputerScope; **ISRAEL'S** Beyond Windows, Computerworld Israel, Multimedia, PC World Israel; **ITALY'S** Computerworld Italia, Lotus Magazine, Macworld Italia, Networking Italia, PC Shopping Italy, PC World Italia; **JAPAN'S** Computerworld Today, Information Systems World, Macworld Japan, Nikkei Personal Computing, SunWorld Japan, Windows World; **KENYA'S** East African Computer News; **KOREA'S** Computerworld Korea, Macworld Korea, PC World Korea; **LATIN AMERICA'S** GamePro; **MALAYSIA'S** Computerworld Malaysia, PC World Malaysia; **MEXICO'S** Compu Edicion, Compu Manufactura, Computacion/Punto de Venta, Computerworld Mexico, MacWorld, Mundo Unix, PC World, Windows; **THE NETHERLANDS'** Computer! Totaal, Computable (CW), LAN Magazine, Lotus Magazine, MacWorld; **NEW ZEALAND'S** Computer Buyer, Computerworld New Zealand, Network World, New Zealand PC World; **NIGERIA'S** PC World Africa; **NORWAY'S** Computerworld Norge, Lotusworld Norge, Macworld Norge, Maxi Data, Networld, PC World Ekspress, PC World Nettverk, PC World Norge, PC World's Produktguide, Publish& Multimedia World, Student Data, Unix World, Windowsworld; **PAKISTAN'S** PC World Pakistan; **PANAMA'S** PC World Panama; **PERU'S** Computerworld Peru, PC World; **PEOPLE'S REPUBLIC OF CHINA'S** China Computerworld, China Infoworld, China PC Info Magazine, Computer Fan, PC World China, Electronics International, Electronics Today/Multimedia World, Electronic Product World, China Network World, Software World Magazine, Telecom Product World; **PHILIPPINES'** Computerworld Philippines, PC Digest (PCW); **POLAND'S** Computerworld Poland, Computerworld Special Report, Networld, PC World/Komputer, Sunworld; **PORTUGAL'S** Cerebro/PC World, Correio Informatico/Computerworld, MacIn; **ROMANIA'S** Computerworld, PC World, Telecom Romania; **RUSSIA'S** Computerworld-Moscow, Mir - PK (PCW), Sety (Networks); **SINGAPORE'S** Computerworld Southeast Asia, PC World Singapore; **SLOVENIA'S** Monitor Magazine; **SOUTH AFRICA'S** Computer Mail (CIO),Computing S.A.,Network World S.A., Software World; **SPAIN'S** Advanced Systems, Amiga World, Computerworld Espana, Communicaciones World, Macworld Espana, NeXTWORLD, Super Juegos Magazine (GamePro), PC World Espana, Publish; **SWEDEN'S** Attack, ComputerSweden, Corporate Computing, Macworld, Mikrodatorn, Natverk & Kommunikation, PC World, CAP & Design, Datalngenjoren, Maxi Data,Windows World; **SWITZERLAND'S** Computerworld Schweiz, Macworld Schweiz, PC Tip; **TAIWAN'S** Computerworld Taiwan, PC World Taiwan; **THAILAND'S** Thai Computerworld; **TURKEY'S** Computerworld Monitor, Macworld Turkiye, PC World Turkiye; **UKRAINE'S** Computerworld, Computers+Software Magazine; **UNITED KINGDOM'S** Computing /Computerworld, Connexion/Network World, Lotus Magazine, Macworld, Open Computing/Sunworld; **UNITED STATES'** Advanced Systems, AmigaWorld, Cable in the Classroom, CD Review, CIO, Computerworld, Computerworld Client/Server Journal, Digital Video, DOS World, Electronic Entertainment Magazine (E2), Federal Computer Week, Game Hits, GamePro, IDG Books, Infoworld, Laser Event, Macworld, Maximize, Multimedia World, Network World, PC Letter, PC World, Publish, SWATPro, Video Event; **URUGUAY'S** PC World Uruguay; **VENEZUELA'S** Computerworld Venezuela, PC World; **VIETNAM'S** PC World Vietnam.

About the Author

David Angell has written 13 books for most major computer book publishers. His most recent include *Mosaic For Dummies*, *The Internet Business Companion*, and *The Instant Internet Guide*. He is a principal in angell.com, an ISDN and Internet consulting firm based in Boston, Massachusetts. For more information, contact David via e-mail at dangell@angell.com.

Credits

Group Publisher and Vice President
Christopher J. Williams

Associate Publisher
Amorette Pedersen

Editorial Director
Anne Marie Walker

Director of Production
Beth A. Roberts

Project Editor
Jim Markham

Manuscript Editor
J.W. Olsen

Technical Reviewer
Anita Freeman

Composition and Layout
Ronnie K. Bucci

Proofreader
C^2 Editorial Services

Indexer
Liz Cunningham

Book Design
University Graphics

Cover Design
Kavish + Kavish

Acknowledgments

My journey to digital enlightenment via ISDN is the result of the generous help and support of many people along the way. These people created a thousand points of light that helped illuminate the path to ISDN conversion.

First and foremost among the angels that helped me stay on the path is Anita Freeman at Pacific Bell. While her official role was technical reviewer, she went way beyond that limited definition. Anita is a world-class ISDN guru that not only knows her stuff but can explain it in plain English to the unenlightened. Also, thanks to Brent Reid at Pacific Bell who fielded many of my questions because he was unfortunate enough to answer the telephone.

Another angel in this project was Karen FitzGerald, Director of ISDN Product & Project Management at Bellcore. She came through at critical times in the life of this project. Jonathan Rawle, Dexter Bachelder at Media Forum, and Jim Burns at Media Solutions, all earned their wings by coming to the rescue at a key point in this project. Last but not least, a warm thanks to the greatest Angell of all, my wife, Joanne.

I'd also like to thank Jerry Olsen for his diligent work as manuscript editor. Thanks to Amy Pedersen, Ann Marie Walker, and Jim Markham at IDG Books for their support during this project.

Finally, a special thanks to the following people for providing support during this project. Axel Leichum at Siemens Stromberg-Carlson; Bob Larribeau at the California ISDN User's Group; Chuck Cederholm at IBM; Donna Loughlin at NetManage; Ed West at ISDN*tek; Edward Hickey at Ameritech; Janice McCoy at the Massachusetts Department of Utilities; Jeremy Goldstein at PicturePhone Direct; Jerry Hopkins at Bell Atlantic; Jim Bergman and Christopher Callendar at PSI; Jon Jackson, Mike McCoy, and Angie Windheim at Intel; Julie Thometz and Jim Lavalle at Digi International; Ken Branson and Royce Hazard-Leonards at Bellcore; Leslie Conway and Everett Brooks at ADTRAN; Lori Lux and Wayne Luu at Combinet; Maribel Lopez Howard at IDC; Mark Goodreau at PictureTel; Mary Campbell and Anthony Antonuccio at Vivo Software; Michael Newman at The Internet Access Company; Norton Lovld at Tone Commander Systems; Pearle Merriner at the Corporation for Open Systems; Rick Yorra, Christine Alessi, and Donna Robertson at NYNEX; Sandy Golding and Dean Sharp at Motorola; Sara Caswell at the National ISDN User Forum; Shawn Glynn at AT&T; Sid Sung at ATI; and Theresa McCarthy and Brian Atkins at ADAK Communications.

The publisher would like to give special thanks to Patrick McGovern, without whom this book would not have been possible.

Contents at a Glance

Cartoons at a Glance

by Rich Tennant

page 109

page 1

page xxvii

page 222

page 69

page 217

page 57

page 157

page 172

page 132

Table of Contents

Introduction

While you're clunking around cyberspace using a modem over a plain old telephone line, telephone companies have been building a new digital telecommunication network called Integrated Services Digital Network — ISDN for short. Digital communication isn't new. What is new is that rest of us can now afford to tap into high-speed digital communication for as little as $15 per month. Those same telephone wires you use now for regular telephone service are the same conduit for ISDN.

What ISDN delivers is data communication at speeds eight to 10 times faster than today's modem — even without any compression. With compression, ISDN can transmit data at speeds up to 512 Kbps. Imagine cruising the Internet at speeds that make the World Wide Web come alive, telecommuting in the fast lane, or grabbing multimedia files in a flash.

But wait, there's more! ISDN is like a Swiss army knife. You can use it to conduct virtual meetings via desktop video conferencing and transform your telephone into a sophisticated business telephone set. You can even do two things at once. For example, you can be talking to someone on the telephone while surfing the World Wide Web — both on a single line. ISDN changes everything, so it's time to get rewired for your digital future.

About this Book

ISDN For Dummies makes your conversion from the bucolic world of analog communication to digital communication via ISDN a smooth one. You'll find that this guide serves two distinct purposes. The first is to make you an educated consumer of ISDN services. The second is to give you hands-on experience working with ISDN to perform a variety of real-world applications. *ISDN For Dummies* boils down a stockpot of technical gobbledygook into a light broth of only the essentials you need to master ISDN's powerful capabilities. Using this guide, you'll learn how to:

- Grasp the Zen of ISDN to understand how it works
- Get ISDN service from your telephone company
- Set up your ISDN connection to get the most from it
- Develop your own ISDN implementation game plan
- Equip yourself with the right ISDN equipment
- Telecommute to your office LAN via ISDN
- Surf the Internet and World Wide Web via ISDN
- Conduct video conferences from your desktop via ISDN
- Save money getting and using ISDN
- Understand the jargon of ISDN
- Install your own ISDN equipment

ISDN comes in two flavors: Primary Rate Interface (PRI) or Basic Rate Interface (BRI). The Primary Rate Interface delivers much more capacity, but it costs hundreds of dollars per month — clearly beyond the reach of most individuals and small businesses. This book focuses on working with the Basic Rate Interface, which is the affordable ISDN service.

How to Use This Book

Get started with understanding ISDN essentials by reading the four chapters in Part I: Getting Started with ISDN. All digital roads start from this part. From there you can explore the specific ISDN applications presented in the rest of the book.

Who Are You?

In writing this book, I make the following assumptions about who you are and what you are seeking from ISDN.

- You're definitely not a dummy. The fact that you're holding this book moves you to the head of the class.
- You've heard about ISDN but don't have a clue what it is all about.
- You're using a PC running Microsoft Windows and you know your way around both. This means you know how to navigate Windows and insert an adapter board in your PC.
- You're searching for a faster ride when you surf the World Wide Web.
- You want to telecommute by connecting to your office network from home so you can work in your bunny slippers.
- You're tired of waiting so long to download and upload files.
- You're always on the lookout for new ways to gain a competitive advantage in today's business environment.
- You'd like to find new ways, such as desktop video conferencing, to extend your reach so you can collaborate with others without added travel.

How This Book Is Organized

ISDN for Dummies has five parts that present the material in a steady progression of what you need to know at the time you need to know it. After building a foundation of ISDN fundamentals, the book goes on to working with specific ISDN applications, such as remote access and desktop video conferencing. Each part is self contained but also interconnected to every other part, which is the natural order of ISDN.

Part I: Getting Started with ISDN

Part I lays the foundation for working and playing in the ISDN realm. In concise terms, you get the essentials to become an educated consumer and user of ISDN. It explains the benefits and pitfalls of ISDN, and how it works. You learn how to order and configure your ISDN connection, as well as what equipment you need for each type of ISDN application. Tying it all together, you'll learn the essentials for creating your own blueprint for getting the most from your ISDN service.

Part II: NT1s, NT1 Plus Devices, and ISDN Telephones

ISDN is a network managed by the telephone company and delivered to your doorstep via a regular telephone line. Part II explains the first piece of equipment you need to deal with at your end of the network — the network termination device. This device defines your premises as a site on the global ISDN network. You learn what your network termination options are, how they work, and which one to use for what type of ISDN uses. Part II also explains which device to use if you want to connect your existing telephone, fax, or modem to an ISDN line, or you want to use only ISDN-ready devices. This part also presents an overview of ISDN's powerful voice communication features. These call management features let you transform a telephone into a sophisticated telephone system.

Part III: Remote Access via ISDN

Remote access is the bread and butter of ISDN service. In Part III you learn to harness ISDN for remote access, including how to connect your PC to an office LAN for telecommuting, surfing the Internet, or connecting to any online service or individual computer. This part explains the maze of remote access options and sorts through them to give you direction in choosing the right solution. It guides you through developing a plan that gives you the most flexibility in your remote access capabilities.

Part IV: Face to Face via Desktop Video Conferencing

This part exposes you to one of the most exciting ISDN applications, desktop video conferencing. For as little as $1000 you can collaborate with geographically dispersed people via real-time video and share applications. Desktop video conferencing systems use a video camera, video capture board, ISDN adapter, and software to turn your PC into a visual communication medium. You'll learn how video conferencing can increase your productivity, and cut your travel and meeting costs. To help you get a feel for this visual medium, you'll experience hands-on working with several desktop video conferencing systems.

Part V: The Part of Tens

As its name implies, Part V provides you with tens upon tens of valuable resource nuggets accumulated during the writing of this book. Much of this information enhances and supports topics covered earlier in this book. Part V includes extensive references to leading ISDN equipment vendors, ISDN Internet service providers, ISDN resources, and more. You'll also find information on ways to save money getting ISDN and get a bird's eye view of the steps needed to get up and running with ISDN.

Appendices

As a bonus, *ISDN For Dummies* includes a tariff summary of what ISDN costs from the major telephone companies in the U.S. Check out Appendix A to see what your telephone company charges for BRI ISDN service. But wait there's more! You'll also get instructions for do-it-yourself ISDN wiring. ISDN uses the same wiring used for regular telephones so why pay a telephone company installer $50-$75 an hour.

Glossary

Finally, the glossary defines all the lingo and jargon of ISDN that you're likely to come across as you walk toward the digital enlightenment of ISDN.

Icons Used in This Book

Several icons are used throughout this book. These icons illuminate important, useful, or just interesting tidbits of knowledge. Here's what they look like and what they denote.

This icon identifies interesting, but nonessential information. I use this icon to provide you with the sort of background information that people who ask "why?" need. You can skip them if you like.

If you're looking for tips and tricks that will save you aggravation and money, here's where you'll find them.

Watch out! ISDN may be the road to digital enlightenment, but the road has numerous pitfalls. This icon points out these problems, and tells you how to avoid them or how to solve them.

This icon marks a special point of interest or supplementary information.

These icons mark tales of my own ISDN experiences and other ISDN-related stories that make interesting reading.

Where to Go from Here

That's all you need to know to get started on your journey to digital enlightenment via ISDN. *ISDN For Dummies* will be your guide as you walk toward the digital light. Use it to illuminate what ISDN is all about and tell you how to harness its digital power. Let's get started with the first step. . . .

Foreword

by Bob Metcalfe

Integrated Services Digital Network (ISDN) is your on ramp to, if you will excuse the expression, the Information Superhighway. And David Angell's book, *ISDN For Dummies* — the one you are holding in your hands right now — is your on ramp to ISDN.

Yes, we know, hearing the term Information Superhighway again is probably more than you can stand. We know it's too long, grossly overused, not nearly a perfect analogy, and when you hear it from a politician, you had better keep your hand on your wallet. Remember, Information Superhighway is what our politicians came up with the last time they tried to shorten the term National Information Infrastructure without calling it Cyberspace.

Since you can look forward to hearing about the Information Super*hype*way again and again for the rest of your sentient life, the least we can do is try shortening it — the term, not your life. Perhaps like this:

National Information Infrastructure
Information Superhighway
Information Highway
Info Highway
Cyberspace
Info Way
I-Way
Iway

But no matter what you call it, ISDN *is* the on-ramp technology for bringing millions of individuals, small businesses, and telecommuters onto the Iway. ISDN *is* small-office, home-office (SOHO) networking. ISDN is replacing what we used to do poorly with those unreliable, whistling and crackling analog modems over voice telephone lines.

And yes, we know, ISDN is not available everywhere, *yet*; isn't easy enough to use, *yet;* and isn't as cheap as it should be in many places, *yet. ISDN For Dummies* is about all that. It's also about where to find ISDN, how to overcome its difficulties, how to get that next factor of ten in speed and convenience for cruising cyberspace with the top down and the wind blowing in your hair.

Dr. Robert M. Metcalfe invented Ethernet at the Xerox Palo Alto Research Center in 1973, founded the Fortune 500 computer networking company, 3Com Corporation, in 1979, and is now an InfoWorld *magazine columnist covering developments along, if you will excuse the expression, the Information Superhighway.*

Part I
Getting Started with ISDN

"NO, NO! YOU HIGH-5 THEM ON THE HAND! THE HAND! NOT THE FACE!"

In This Part...

The foundation is laid for working and playing in the digital wonderland of ISDN. In concise terms, you get the essentials to become an educated consumer and user of ISDN. You learn what ISDN delivers, how it works, how to connect to it, and what equipment you need. For the grand finale, you learn how to tie it all together into a strategy for getting the most from your ISDN connection. It's a lot to cover, but once you complete this boot camp of ISDN communications, you're ready to roll up your sleeves and start with specific ISDN applications.

Chapter 1

We're Not in Analog Anymore, Toto

In This Chapter

- How we live in the bucolic analog communications era
- Why ISDN is destined to be your digital communications future
- What ISDN can do for you
- Why ISDN is not plug and play — yet

Turn on your answering machine, turn off your modem, and take a little time to become acquainted with your digital communications future. In case you haven't noticed, you live in the digital age. Images, sounds, videos, and just about any other form of information are widely available in digital form. Yet, the network millions of us use to transmit this data continues to clunk along using technology that has been around for almost 100 years — the telephone system. That is, until now. You're about to enter the age of digital communications via ISDN — Integrated Services Digital Network.

Life in the Bucolic Analog Communications Era

The ubiquitous telephone system all of us use today transmits the human voice using the analog device known as the telephone. This analog system transforms sound into electrical signals. The telephone network contains amplifiers along the way to boost those signals for long distances. This voice-based system is called *POTS* in telecommunications circles, which stands for *plain old telephone service.*

With the advent of computers, we started to use the telephone system as a data communication link to other PCs, networks, online services, and more recently the Internet. Computers now impact the makeup of our communications over the telephone system. The modem is the current staple for data communications over the analog telephone system for most PC users. The term modem is a contraction of its function, that is, modulator and demodulator. The modulator function changes digital data bits into sound waves to fit through phone lines, while the demodulator function changes the sound waves back into digital data. Over time, modems have to become faster, with 9.6 Kbps and 14.4 Kbps modems in common use today. At the upper reaches of modem technology are the new 28.8 Kbps modems. However, today's modems are coming up against the data speed limit of POTS.

In the analog world, you use the telephone for voice, a fax device to send documents, and a modem to connect to cyberspace. If you have only one POTS line, you can perform only one task at a time. To conduct multiple tasks at the same time, you must have a separate analog line for each device. Figure 1-1 shows a typical POTS line setup for a small business or work-at-home scenario. This bucolic analog communication age in which we all grew up is in its twilight years.

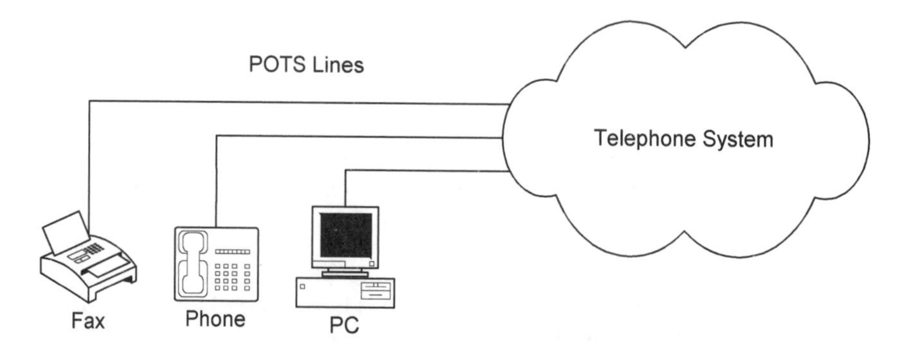

Figure 1-1: The typical analog layout for working with a telephone, fax, and a PC with the capability to use each device at the same time requires a separate line for each device.

GUI Gets Stuck in the Modem Bottleneck

The world of PC computing has changed from a barren, text-only environment to a rich, graphical, and multimedia realm. Microsoft Windows dominates the PC platform, and Windows application file sizes now involve megabytes instead of kilobytes. As software gets easier and more powerful, file sizes continue to grow bigger and bigger. This migration to easier and more powerful computing comes at a price. The price is the need for more computer processing power and more capacity for transporting data. The effect on PC hardware is a demand for faster CPUs, faster graphics adapter cards, faster drives, and so on.

Squeezing large amounts of information through a modem and POTS telephone line just doesn't cut it in the modern *GUI* (*graphical user interface*) and multimedia computing environment. A good example of the modem bottleneck is surfing the World Wide Web. This popular Internet tool lets people work on the Internet using a Windows browser like Mosaic or Netscape to interact with information in a hypertext and multimedia context, as shown in Figure 1-2. Vast amounts of data are transferred to the user, including image, sound, and video files. This kind of data volume simply goes beyond current practical modem and POTS capabilities.

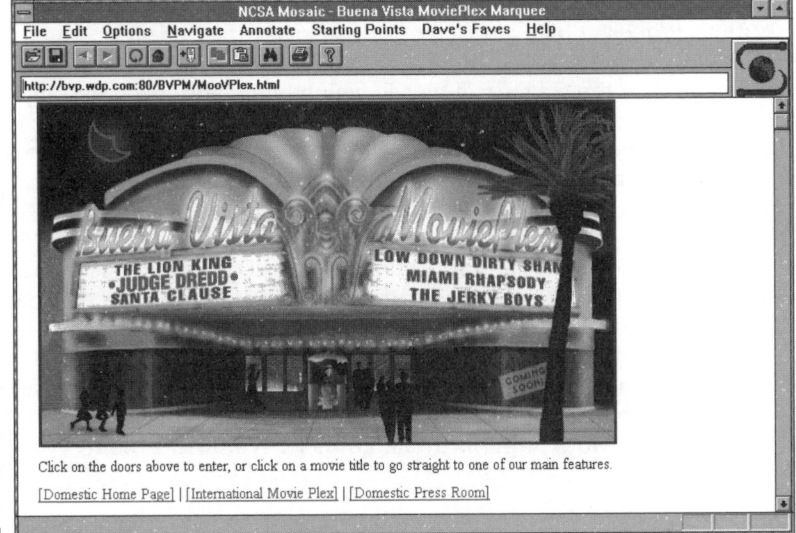

Figure 1-2: A typical, data-intensive view of the Internet through a Mosaic for Windows user's eyes.

Digital communications primer

Digital means data based on the binary number system, which uses only 0 and 1 to represent all information. By manipulating these electronic on and off switches, computers manage information in any form. Digital communication is the exchange of information sent in binary form instead of a series of analog signals. The common measurement of data transmission speed is *bits per second (bps).* The term bit is a contraction of binary digit, the smallest unit of digital information, of which eight bits make one byte. A byte is the equivalent of a single character. *Kbps* stands for *kilobits per second,* which in a computer setting is 1024 bits per second. Mbps stands for megabits per second, which is 1,048,576 (1024 times 1024) bits per second. MB stands for megabytes, which is 1,048,576 bytes. This megabytes measurement is typically used in conjunction with files and data storage.

And the Bandwidth Played on

In popular telecommunications and networking lingo, *bandwidth* means data transmission capacity. The more the bandwidth supported by a channel, the more data it can pass through in a given amount of time.

All the whiz-bang technology of graphical and multimedia computing demands more bandwidth to function at a reasonable pace. Sure, you can send a huge file over an analog line using a 9.6 Kbps modem, but it's not viable because of the time and connection charges. That same file transported through a digital line at 128 Kbps gets transferred in a fraction of the time. It doesn't take a rocket scientist to see the benefits.

It's not just the transfer of files in which bandwidth becomes important. The bandwidth of a connection also affects the data transfer rate for real-time images, sounds, or videos from a host computer to your PC. The bottom line is that the more bandwidth a communication link has, the more sophisticated the applications that it can support, such as video conferencing.

For most individuals and small businesses, POTS communications — the lowest level in the data communications food chain — was the only game in town. In the world of data communications, the rule has been the greater the bandwidth, the more expensive the line. Before ISDN, the only option for digital communications has been leasing dedicated digital lines. This is an expensive proposition that costs hundreds of dollars per month. Rules change, and now digital bandwidth power is available via ISDN at prices the rest of us can afford.

Walk toward the Digital Light

Integrated Services Digital Network (ISDN) is the next generation of communication service offered by telephone companies, and is widely available right now. What ISDN brings is affordable, high-speed digital communications — up to 128 Kbps — to millions of individuals and small businesses for the first time. The telephone company pipes ISDN service into your home or office just like POTS. What ISDN delivers is nothing short of a communications revolution.

ISDN Benefits in a Nutshell

OK, enough of the vision; let's look at the specifics of what ISDN offers in practical terms. ISDN is an integrated voice and data network that offers an impressive collection of benefits.

Bandwidth isn't the girth of a heavy-metal band

The capacity of any data communications line is a function of its range of frequencies for carrying information, which is called *passband*. Bandwidth is the width of the passband, the difference between the highest and lowest frequencies in a given range. For example, the human voice produces sounds in the passband of 50 Hz to 15,000 Hz (15 kilohertz or 15 kHz) that translates to a bandwidth of 14,950 Hz (15,000 Hz minus 50 Hz). In this case, Hertz (Hz) is not a car rental agency, but a unit of frequency measurement for a waveform cycle.

- ISDN handles all types of information. It's one network that can move many different types of information, including voice, text, images, sounds, and video.
- ISDN delivers data at speeds up to 128 Kbps without any compression, which translates to eight to 10 times the speed of today's 14.4 Kbps modems. With compression, ISDN data speeds can reach 512 Kbps.
- ISDN is affordable and cost effective. Basic ISDN service costs an average of $35 to $50 per month, and a single ISDN line can replace the functions of multiple POTS lines.
- ISDN uses the same telephone wiring as POTS. You don't need new wiring to handle ISDN service in most homes and businesses.
- ISDN lets you perform more than one communication task at the same time. You can speak on the telephone to someone while surfing the Internet using a single ISDN line.
- ISDN offers impressive call-management features for voice communications like Caller ID, call forwarding, call conferencing, and multiple incoming calls on the same line.
- ISDN lets you connect up to eight devices on a single line. You can get multiple telephone numbers and call appearances added to a single ISDN line to handle an expanded volume of calls.
- ISDN is part of the ubiquitous analog telephone system, so you can communicate with ISDN users as well as people still connected to the analog world.
- ISDN lets you connect and use your existing analog devices like telephones, faxes, and modems, yet allows you to switch to data transmission overdrive when you need it.

- ISDN lets you connect credit card readers or other devices to communicate via X.25 packet networks. Small businesses can use ISDN as an inexpensive alternative to expensive leased lines for credit card authorizations.
- ISDN is an on-demand service that you use only when you need it. It offers an affordable alternative to expensive, dedicated digital lines that cost hundreds of dollars per month.
- ISDN offers greater accuracy and connection stability. AT&T estimates a POTS connection has a reliability rate of 75 percent; a digital connection, up to 100 percent.

Where Did ISDN Come from?

ISDN's implementation by telephone companies has taken a long time to come of age. It has had several false starts along the way, and ISDN's march to the mainstream has left a trail littered with the disillusioned. Many early adapters became followers of the ISDN acronym, It Still Does Nothing. But ISDN's full implementation was inevitable because the march toward a digital telephone network has been going on for years.

Telephone companies have been converting from analog to digital for several decades. The digital system uses computers to perform switching tasks that route calls. The introduction of digital switches allowed the telephone network to operate more efficiently and allowed telephone companies to offer new communication services to customers. Many of these services you're already familiar with, such as call waiting, call forwarding, and conference calling.

At the same time that the telephone companies were going digital, the interface to you and the rest of the world remained POTS, to maintain compatibility with the huge consumer analog equipment infrastructure as well as for other political and economic reasons. It's this last link between the telephone company and customers that finally delivers ISDN to the rest of us. By the end of 1996, most telephone companies expect to have 85% to 95% of their lines capable of handling ISDN. Right now, most large metropolitan areas in the U.S. are ISDN ready. However, ISDN deployment varies among different telephone companies, as explained in Chapter 2.

What ISDN Can Do for You

Not only can ISDN deliver dramatic improvements in the types of communications you're already doing using POTS, it's also a powerful conduit for all kinds of new voice and data services that just aren't feasible via POTS. Many of these new uses such as personal video conferencing, are already available. Expect to see many more.

Are You a Candidate for ISDN?

The combined voice and data capabilities of ISDN offer a broad range of applications for all kinds of people, including telecommuters, small office/home office (SOHO) workers, Internet surfers, consultants, contract workers, business executives, researchers, teachers, and customer service representatives. Organizations of all sizes can use the power of ISDN to conduct business more efficiently. Are you a candidate for ISDN service? You are if you:

- Work at home as a telecommuter and must connect to your company's network.
- Operate a home-based business.
- Use multiple analog telephone lines for a telephone, fax, and a modem.
- Send and receive large files, such as image, video, or sound files.
- Depend on voice communications via the telephone for conducting your business.
- Make regular connections to the Internet or other online services.
- Collaborate with geographically dispersed people.
- Frequently travel to attend meetings.

The following sections cover the key uses of ISDN today. However, keep in mind that ISDN is not a static network system, and new applications are sure to come. For example, ISDN is an excellent conduit for alarm systems, utility meter reading, energy management, and other services that operate in the background to monitor your home or business.

Telecommuting (or Working in Your Bunny Slippers)

Telecommuters are employees who work from home part or full time during normal business hours. Telecommuting is a rapidly-emerging trend as companies downsize, people seek different working options, and environmental and social changes converge.

According to LINK Resources, a market research organization, telecommuters number 7.6 million and their numbers are growing at an annual rate of 15%. According to the Gartner Group, a corporate consulting firm, more than 80% of all organizations will have at least half of their staff engaged in telecommuting by 1999. ■

ISDN offers an affordable, high-speed connection for telecommuters to connect to their office computers. ISDN's bandwidth and sophisticated voice and data integration allow people to work off site and yet interact with other people as if they were on site. A study conducted at West Virginia University found that transmission speeds of 80 Kbps made users feel like they were actually connected to a local area network (LAN). ISDN delivers 128 Kbps without any compression. Figure 1-3 shows a typical telecommuter ISDN configuration that integrates voice and data services.

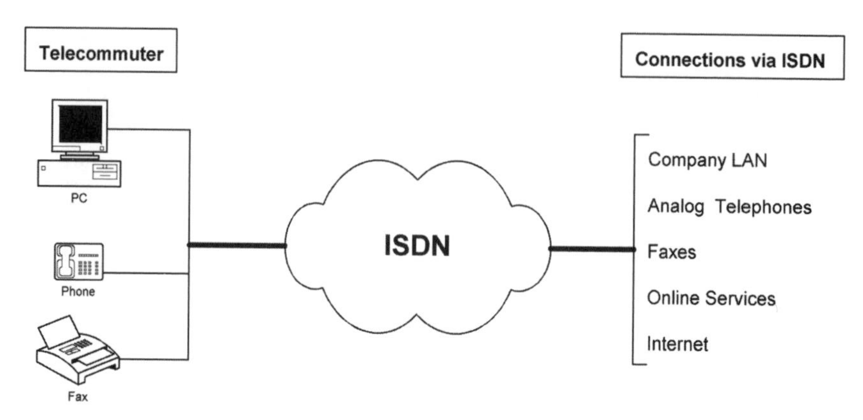

Figure 1-3: The typical telecommuter can use multiple ISDN services to interact with people at or outside the office.

Cruising Cyberspace with the Top Down

Cruising cyberspace with ISDN is like driving on the open road with the top down and the wind blowing in your hair. ISDN offers reliable, high-speed access to the Internet, online services, and databases. The impact of ISDN promises to make cyberspace a more exciting, cyberdelic realm of intensified sensory perception. The greatly-increased data transfer speeds of ISDN let information providers incorporate a wide range of graphics — photos, drawings, illustrations, charts, and diagrams — few of which are practical at today's modem speeds and reliability.

ISDN offers an express lane for working and playing on the Net. A growing number of Internet service providers offer ISDN connections at affordable prices. You can connect a small network to allow multiple users to access the Net for a variety of everyday tasks from e-mail to file transfer. An ISDN connection moves information at speeds that let the popular World Wide Web come alive on your desktop.

One of the first things a modem user discovers when surfing the World Wide Web is how slowly information travels. An ISDN connection dramatically speeds up your Web surfing adventures. Web documents and files are downloaded swiftly. The reliability of an ISDN connection is dramatically improved compared to an analog connection. ■

The near-term benefits of ISDN for users of such online services as CompuServe, America Online, and PRODIGY will be the same as for Internet users. You'll be able to perform all kinds of daily tasks at unprecedented speeds using an ISDN connection. ISDN also allows information providers of all kinds to create sophisticated multimedia databases to deliver information to subscribers. For example, a real estate database can include pictures, layouts, and even videos that people can browse online.

Adding Digital Power to Your Voice Calls

ISDN brings a whole new generation of voice and call management services to your telephones. The power of ISDN lets you handle multiple telephone activities on a single line.

ISDN service working with your existing analog equipment lets you add a cornucopia of call management features. Many of these telephone services are already familiar to you, including call holding, call waiting, and call forwarding. What is so powerful about these features in ISDN is that you can use a whole bunch of them at one time across multiple telephone numbers from a single ISDN line.

Collaborating beyond the Bounds of Geography

ISDN is the communication glue that allows geographically dispersed people to work together in real time. ISDN offers the bandwidth for a variety of virtual workgroups ranging from a simple conference call with computer screen sharing to a real-time video conference from the desktop.

ISDN can support users talking while jointly viewing and editing the same file. Screen sharing, commonly referred to as a *whiteboard* session, allows two or more users to work interactively with voice and data using a single ISDN line.

A more powerful form of desktop collaboration involves adding a video conferencing element to a whiteboard session. As a business tool, video conferencing delivers a low-cost alternative to travel as well as an effective strategic tool for getting things done faster. ISDN is an efficient transmission medium for video conferencing because its digital basis provides virtually error-free transmission for high-quality images. ISDN permits simultaneous voice and data transmission over the same line. A new generation of personal video conferencing makes this an affordable option. Figure 1-4 shows how a video conference appears on a PC screen.

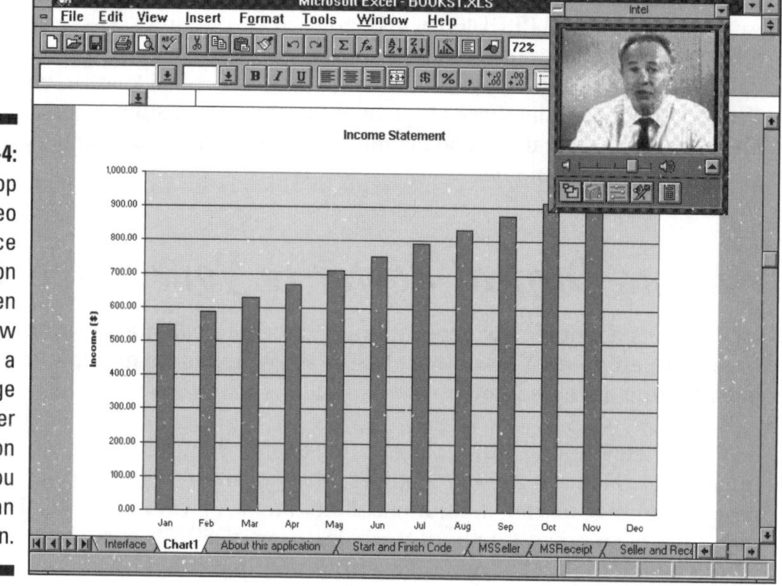

Figure 1-4: A desktop video conference appears on your screen as a window displaying a video image of the other person while you share an application.

Moving Files at Lightning Speed

The size of files grows almost exponentially as computer applications become more sophisticated. ISDN lets you move all kinds of files quickly and reliably between PCs or via the Internet. Desktop publishers, multimedia developers, software companies, and a host of others can deliver their software products economically and efficiently via ISDN. Using standard PC communications software at both ends of a connection, you can transfer files in the same way you do with a modem, except at up to eight times the speed of a 14.4 Kbps modem.

Conducting Transactions in the Cashless Society

ISDN allows even the smallest of businesses to use automated credit and debit card authorizations. Using an inexpensive card reader and ISDN connection, a business can receive an authorization in seconds. These authorizations can occur in the background as you use the same ISDN line for other forms of communications. With ISDN a business doesn't need expensive dedicated lines or slow modem dialup services to receive authorizations.

ISDN Is for On-Demand Digital Communications

ISDN brings affordable, on-demand digital communications to small businesses and individuals. The key word here is on-demand as opposed to dedicated. Dedicated digital telephone lines have been around for a long time. They're leased lines that operate 24 hours each day for a fixed monthly rate, which is usually in the hundreds of dollars. There is no usage charge associated with the line.

Because ISDN is an on-demand system, it's treated like telephone communications in terms of charges. You make a call and the telephone company charges you. If you make a long distance call, the long distance telephone company charges you for the call.

This difference is crucial to understanding the limits of ISDN service. ISDN is fine for making calls to perform specific tasks on an as-needed basis or for accepting incoming calls, such as for a computer bulletin board system. But its not an appropriate solution if you want to set up a World Wide Web server on the Internet. It's not economical to use an ISDN line as a dedicated connection 24 hours per day. That's where the economics of a dedicated leased line come into play.

ISDN Isn't Plug and Play — Yet

As you enter the era of digital communication, understand that ISDN is a new technology and as such is going through growing pains. ISDN is a complex system that integrates telecommunications and computers. It's composed of disparate, yet interdependent players, including telephone companies, hardware and software vendors, governmental agencies, and others.

The good news is that many of ISDN's early problems have for the most part been fixed. Those problems included interoperability between telephone companies, lack of universal service, and complex installation procedures. Telephone companies have made great strides in addressing these problems. New tariffs are in place that make ISDN attractive to small businesses, telecommuters, and others. ISDN equipment is getting better and more affordable.

The bad news is that ISDN is still not a yellow brick road. Stretches of the ISDN network are still under construction, incompatibilities remain, and equipment capabilities and costs vary widely. But it's getting better, and the rewards are worth the effort. The following sections present an overview of the main trouble spots in ISDN service today.

Relax, pointing out ISDN's problem areas is part of your digital enlightenment program. As you deal with specific ISDN tasks, this guide will help you navigate safely through the rocky shoals. ■

Inconsistent ISDN Service

Your local telephone company delivers ISDN to homes and businesses. The U.S. has seven regional telephone companies, such as Pacific Telesis, NYNEX, and Bell South. Each of these companies operates independently, which results in uneven ISDN service availability, pricing, and service. Some telephone companies are aggressively trying to make ISDN easy to use. Others continue to be sluggish in ISDN implementation. The result is that your ISDN experience can be good or bad depending on your telephone company.

Complex ISDN Pricing

ISDN pricing, called *tariffs* in telephone company lingo, comes together through the application of complex cost allocation and recovery rules established by both federal and state regulators. Thus, ISDN pricing varies from one telephone company to another and from one state to another. When you subscribe to ISDN service, your final price depends on a variety of configuration options offered by your telephone company. Determining the optimal configuration for your ISDN service is important because telephone companies often charge hefty installation charges (typically over $100) and other fees for line configuration changes. Adding to the confusion, telephone companies also charge different rates for different types of traffic going over your ISDN connection. While progress is being made in standardizing prices and services, figuring out your options and costs remains one of the most difficult parts of establishing an ISDN connection.

Unfriendly ISDN Setup

The most difficult part of setting up an ISDN connection is configuring your ISDN line for the equipment you use. The gray area between the telephone company ISDN connection and your ISDN equipment can be a great and frustrating divide. To make an ISDN connection work, you must deal with two different players, the telephone company for the ISDN line and ISDN equipment vendors for the devices at your location. In the past, these two groups have not cooperated well. Fortunately, this state of affairs is changing as telephone companies and ISDN equipment vendors work together to make setting up an ISDN connection easier. One way is through ISDN ordering codes that act as configuration templates. The ISDN customer simply tells the telephone company representative a code supplied by the equipment vendor and the telephone company can automatically configure the ISDN line.

Incompatible and Complicated ISDN Equipment

ISDN vendors lack the standards in certain ISDN applications to ensure the interoperability of equipment. However, this is changing as standards take hold in various ISDN applications, such as video conferencing and Internet access. Beyond incompatibility, many ISDN equipment vendors are not end-user oriented in terms of marketing and supporting ISDN products. The ISDN equipment market has more than its share of poorly designed devices, incompatible systems, and poor documentation. Adding to the problem, ISDN supports multiple applications, each with its own requirements and equipment that end users must assemble and integrate into a total ISDN solution.

Moving on . . .

At this point, you've peeked into the digital future of ISDN communications. You've been introduced to the promise and pitfalls that await you on your journey toward digital enlightenment via ISDN. The next chapter explains the fundamentals of how ISDN operates as a system from the telephone company grid to your side of an ISDN connection.

Chapter 2

The Zen of ISDN

In This Chapter

- How the telephone company side of ISDN works
- How protocols and standards impact your ISDN connection
- What the two ISDN access options are
- What the ISDN functional devices are
- What the reference points of an ISDN connection are

Before you jump into the mechanics of ISDN, you need to understand the guiding principles that define the operation of ISDN. These fundamentals include the systems and rules that make up ISDN and encompass the telephone network side as well as your side of the connection. Think of these fundamentals as the Zen of ISDN, because understanding them gives you valuable insights for working with ISDN. Continue on, Grasshopper.

The Telephone Company Side of ISDN

The integration of the local and long distance telephone companies form one big ISDN network. Your local telephone company brings ISDN to your doorstep via the standard POTS telephone wiring. From this demarcation point, you're in charge of the ISDN connection. It's helpful to think of ISDN as a network that is operated by all the telephone companies. When you establish an ISDN connection, your premises becomes a node on this global network. To manage your ISDN connection, you need to understand the basic operation of the ISDN system from the telephone company side to your ISDN equipment.

A common symbol for any large network that delivers information in a way that is transparent to the users of the network is a *cloud*. ISDN is such a network. You'll see the cloud used in illustrations in this book as well as other ISDN documentation. ■

In the United States, regional telephone companies provide the telephone service. Each company, called a *Regional Bell Operating Company (RBOC)* serves a specific area, as shown in Figure 2-1. These companies are Ameritech, Bell Atlantic, Bell South, NYNEX, Pacific Telesis, Southwestern Bell Corporation, and US West. Another term for RBOCs is *Local Exchange Carrier (LEC)*.

In addition to the RBOCs, there are *independent telephone companies (ITCs)*. All of these companies provide local service and send that phone bill you complain about every month. Long distance telephone companies like AT&T, MCI, and Sprint are called *InterExchange Carriers (IC or IEC)*. They provide service between local telephone companies.

Figure 2-1: Service areas for the Regional Bell Operating Companies in the U.S.

Telephone Company Switching Systems

A *switch* is a general term referring to facilities that route telephone traffic from one destination to another. ISDN is the result of the marriage of computers and telecommunications, called *telephony.* Electronic switching software operated on computers provides the basis for the operation of ISDN. These *digital switches* provide electronic routing for telephone calls, with telephone numbers acting as a routing address system.

The telephone companies operate a hierarchy of switching, depending on the call. At the local level, calls can often be completed within a single switch. Beyond this, there are switches for long distance and regional calls. During a long distance telephone connection, the call starts via your local telephone company, which switches the call to the long distance carrier to route the call to the local telephone company at the other end of the connection, as shown in

Figure 2-2. Digital circuit switches that control the routing also manage other parts of the telephone system, such as determining toll charges and providing call management features like call waiting and call forwarding.

ISDN service is a circuit switching and packet switching system. The term *circuit switching* means that the communications pathway remains fixed for the duration of the call and is unavailable to other users. A circuit switched connection between two users becomes a fixed pathway through the network. When the calling party initiates the call, a set-up procedure tells the network the address of the calling party and sets up the route for the call. The call can be placed for the transmission of data or voice. The term *packet switching* refers to the sending of data in packets that are sent by the most efficient route and then reassembled at their destination.

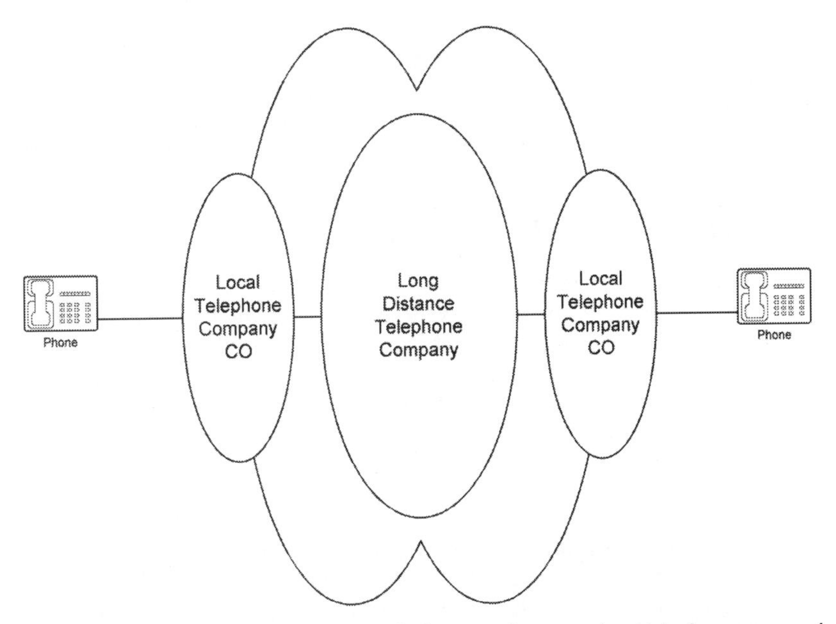

Figure 2-2: The path your long distance call takes goes from your local telephone company's Central Office through the long distance company's network to another local telephone company.

The leading digital circuit switches used by the telephone companies are AT&T's *5ESS* (Electronic Switching System) and Northern Telecom's (NT) *DMS-100* switch. These two computer-based switching platforms use operating system software. The AT&T 5ESS switch uses Custom or National ISDN 1 (NI-1) software and the DMS-100 uses NI-1. The majority of telephone companies use these switches. You see references to them in all ISDN equipment documentation because compatibility between your ISDN equipment and the telephone company's switches is necessary to communicate via ISDN.

How did we get so many telephone companies?

Prior to 1982, AT&T held a monopoly on telephone service in the United States. In that year, the U.S. Government settled an antitrust lawsuit against AT&T, which became the basis for the breakup of AT&T into the seven RBOCs. AT&T retained long distance communication services and communications equipment manufacturing. The 22 Bell operating companies became the seven regional Bell companies that operate today. These RBOCs received control of local switching equipment, and their service areas were divided into approximately 160 *local access and transport areas* (*LATAs*). A LATA usually comprises many different area codes. For example, LATA 1 in California goes from the northern point of the state at the Oregon border south to Monterey. Within LATA 1 are several area codes. There are ten LATAs in the state of California, and hundreds of area codes. RBOCs can only handle calls within a LATA, while long distance telephone companies carry calls across LATAs.

The Central Office

A *Central Office* (*CO*) is a local telephone company facility that houses all of the telephone company switches. The CO is the wiring center or hub for all of the telephone company's subscribers. It's the massive cable plant at the CO that provides the path from the switching device to the customer's premises. All kinds of call management and call routing activities take place at the CO, including local and long distance switching. It's the front line of telephone service from the telephone network to your premises and vice versa. There are about 19,000 Central Offices in the U.S.

Large metropolitan areas usually have a bunch of COs, each serving a specific geographical area. These Central Offices connect to other COs for local calling, or to other switching facilities for long distance calls. COs are the last part of the telephone network to be upgraded to support universal ISDN service.

Local Loops (No, It's Not a Cereal)

The term *local loop* refers to the telephone line comprising the pair of copper wires between your premises and the telephone company's CO. There are more than 200 million local loops in the United States. The wiring for ISDN from the telephone company to your premises is the same copper wires used for POTS but requires different equipment at both ends of the connection. The local loop connects your premises to the CO, which in turn connects you to the ISDN cloud, as shown in Figure 2-3.

As a rule of thumb, the maximum length of a local loop for ISDN is about 18,000 feet using standard POTS wiring without any repeaters. A repeater is any device that amplifies or regenerates the data signal to extend the distance of the transmission. This limitation is one of the main reasons that ISDN service may not be available to your premises. However, repeater technology is being deployed to extend this limit another 18,000 feet. As the telephone system moves from copper wires to fiber optics technology, the current distance restriction will disappear. *Fiber optics* send light beams through thin strands of glass or other transparent material to transfer data. These optical fibers are immune to the electrical interference inherent in POTS.

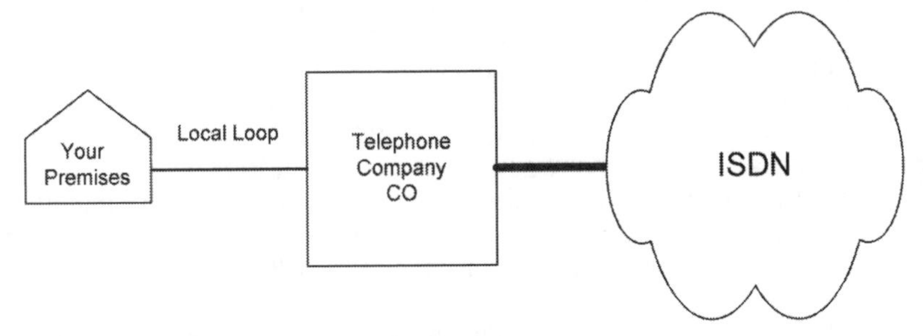

Figure 2-3: The local loop is your connection to the telephone company CO and on to the ISDN cloud.

More Power, Scotty

As you may have noticed, your telephone service remains active when the power goes off. This is because the telephone company provides the electrical power for the analog line from the CO. Unlike a POTS line, an ISDN line is not powered by the telephone company. End users of ISDN are responsible for providing the power to send digital signals through the ISDN lines between their premises and the CO. Power requirements for an ISDN line are minimal and are supplied by a device that plugs into a standard electrical outlet.

In Your Interface

In the lingo of networking, an *access interface* is the physical connection between you and the network that allows you to use the network. Access to ISDN comes in two forms: *Basic Rate Interface* (*BRI*) and *Primary Rate Interface* (*PRI*). These access interfaces deliver digital transmission services via channels. A *channel* is a conduit through which information flows in both directions. The BRI and PRI access interfaces define the channel configuration for the two types of ISDN service.

ISDN Service for the Rest of Us

The Basic Rate Interface (BRI) is the affordable access service that most small businesses and individuals use for ISDN. For BRI service, the telephone company divides the POTS wiring between your premises and the CO into three separate logical channels. The term logical refers to the fact that these three channels are not three separate wires, but are defined by the logic of how the ISDN system operates.

The three channels of a BRI connection include two 64 Kbps B *(Bearer)* channels and one 16 Kbps D *(Data)* channel, as shown in Figure 2-4. This standard BRI configuration is referred to as *2B+D.* The two B channels are the workhorse of an ISDN connection because they deliver the bulk of the digital information that's transmitted. This digital information can include voice, files, video, and sounds. You can use each B channel separately or your ISDN equipment can combine the two B channels to allow data transmission speeds up to 128 Kbps without compression. You also can combine multiple BRI lines to add more bandwidth power. For example, using two BRI lines and the right equipment, you can get transmission rates of 256 Kbps (64 Kbps × 4).

The D channel delivers signaling information that tells the telephone company switches what to do with the stuff that's being delivered via the B channels. This signaling information opens and closes circuit switches to route calls. The 16 Kbps capacity of the D channel is more than enough for ISDN signaling needs, so you can use it to transmit X.25 data. Public data networks, such as the CompuServe network, use X.25 as well as credit card reading devices.

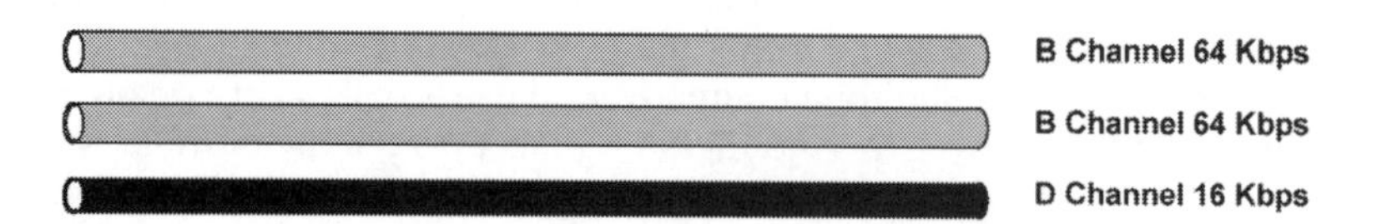

Figure 2-4: The Basic Rate Interface consists of two 64 Kbps B channels and one 16 Kbps D channel.

Keeping Your Signals Straight

The term *in-band signaling* refers to the delivery of the information to operate the communications channel being carried in the same channel as the information. For example, in POTS the signals used to control the network are also carried over the same line that carries your voice.

Out-of-band signaling sends the information that controls the network via a channel separate from the one that delivers the user's information. ISDN uses out-of-band signaling via the D channel. *Signaling System Number 7* (*SS#7*) is a collection of signaling switching standards that allows out-of-band signaling to work across the different switches of the telephone network. This allows ISDN to be a network beyond just your local connection to the telephone company.

However, because local telephone companies are slow to implement out-of-band signaling connections to long distance carriers, many long distance calls combine the signaling into the B channel. This system uses 8 Kbps for in-band signaling. The result is that your data in the B channel is delivered at 56 Kbps or 112 Kbps instead of 64 Kbps or 128 Kbps for the two B channels.

Raw ISDN Power for Deep Pockets

The Primary Rate Interface delivers 23 64 Kbps B channels and one 64 Kbps D channel. PRI service is referred to as *23B+D*. As you might guess, this digital communication power comes with a price tag that is out of reach for small businesses and individuals. Companies installing multiple ISDN lines use PRI as a kind of wholesale delivery of B channels, which they then divide at the premises.

Because of the cost of PRI service and the demands for installing it, working with PRI is beyond the scope of this book. This book deals exclusively with the BRI side of ISDN.

Playing by the ISDN Rules

ISDN is a network that works because of protocols and standards. A *protocol* is a set of rules for data communications, and protocols define the workings of ISDN as a network. A *standard* is a set of detailed technical guidelines that establish uniformity. Protocols and standards permit different equipment and network systems to work together.

To develop and publish universal rules and conventions, many industries and countries have established standards organizations. As a complex, global system, a number of organizations are involved in defining ISDN operation. In the ISDN realm, you frequently see references to protocols and standards defined by these organizations. Table 2-1 describes the main standards organizations that help to bring order to ISDN service and equipment.

Table 2-1: ISDN protocol and standards organizations

Organization	*Description*
Telephone Consultative Committee or Comité Consultatif International Télégraphique et Téléphonique (CCITT)	The organization primarily responsible for producing international ISDN recommendations. This organization has been consolidated into the International Telecommunication Union (ITU), an agency of the United Nations.
ANSI	The American National Standards Institute is the primary standards setting body in the U.S. ANSI plays a significant role in the development of ISDN standards. In the U.S., ANSI defines the operations of the local loop.
North American ISDN User's Forum (NIUF)	The National Institute for Standards and Technology formed the NIUF in conjunction with ISDN industry players to identify ISDN applications and encourage ISDN equipment vendors to develop products of interest to users.
Corporation for Open Systems (COS)	Within the U.S., testing and certification for ISDN is specified and performed by COS.
Federal Communications Commission (FCC)	This federal agency develops and publishes rules and regulations that govern communications and communications equipment in the U.S.
Bell Communications Research (Bellcore)	Formed as the research and development arm of the seven RBOCs in the U.S., Bellcore plays a leading role in defining ISDN standards for the RBOCs. It has developed National ISDN, which is a collection of specifications for uniform implementation of ISDN.

National ISDN 1

After the breakup of AT&T, Bell Communications Research (Bellcore) was formed as the research and development arm of the seven RBOCs in the U.S. Bellcore plays a leading role in defining ISDN standards for the RBOCs. It developed National ISDN 1 (NI-1), a family of specifications to make ISDN equipment and service interoperable. Nearly all major switch manufacturers, telephone companies, and CPE vendors in the U.S. support NI-1. Bellcore has introduced National ISDN 2 and National ISDN 3 specifications to add even more standardization of ISDN services and products.

Functional Devices and Reference Points

ISDN protocols define the several types of devices, as well as the communication between devices, that together comprise the total ISDN connection. The protocols that define the types of devices used in an ISDN connection are called *functional devices*. The protocols that define the communication between functional devices are called *reference points* or *interfaces*. The important thing to remember about functional devices and reference points is that they're protocol definitions. These protocols become embodied in ISDN equipment in various configurations. As a buyer of ISDN equipment, you need to understand functional devices and reference points to evaluate the role of each piece of ISDN equipment. Figure 2-5 shows the complete layout of an ISDN connection's functional devices and reference points — each of which is explained in the following sections.

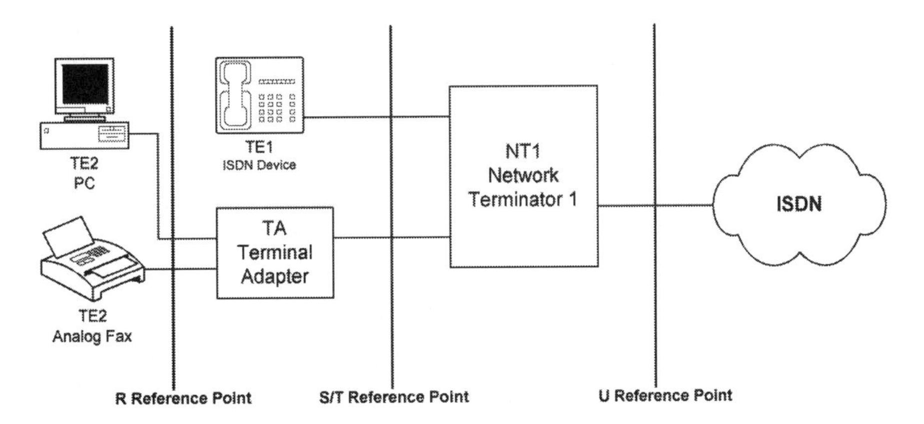

Figure 2-5: ISDN functional devices and reference points are a collection of protocols that define the ISDN connection.

ISDN Functional Devices

Functional devices are definitions for specific tasks performed in an ISDN connection. These functional devices reside on the user side of the connection. Actual ISDN equipment often embodies multiple functional devices, but may include only one. The following sections explain the functional devices in an ISDN connection.

All Digital Roads Lead to Network Termination (NT1)

The *Network Termination 1* (*NT1*) represents the boundary to the ISDN network from the end-user side. NT1 includes the physical and electrical termination functions of ISDN on your premises. Specifically, NT1 is the device that provides an interface between the twisted-pair wires used by the telephone company in the BRI and the eight-wire cables used by ISDN equipment. Because the ISDN line doesn't provide power like the analog line does, NT1 also includes the power function for operating the ISDN line. Each BRI access has only one NT1 device. NT1 can be embodied in a stand-alone device or included in a specific device.

A *Network Termination 2* (*NT2*) device handles switching functions and is embodied in PBXs (Private Branch Exchanges). The NT2 device performs intelligent operations such as switching and concentrating traffic across multiple B channels in a PRI line. Working with PBXs and the Primary Rate Interface is beyond the scope of this book. ■

Terminal Equipment (TE)

Terminal Equipment (*TE*) refers to any end-user device connected to an ISDN line. This is the general class of equipment that covers both ISDN-ready equipment and non-ISDN equipment such as analog telephones, faxes, and modems. The encompassing term for any equipment on the customer side of the ISDN connection is *customer premises equipment* (*CPE*). ISDN can support up to eight pieces of CPE to perform multiple tasks using a single line.

Terminal Equipment 1 (TE1)

Terminal Equipment 1 (*TE1*) refers to ISDN devices that support the standard ISDN interface directly, including digital phones, digital faxes, and integrated voice/data terminal devices. These TE1 devices provide direct access to an ISDN connection without adapters. Currently, most ISDN-ready equipment is too expensive to be practical, but expect to see low-cost ISDN telephones and other devices in the near future.

Terminal Equipment 2 (TE2)

The *Terminal Equipment 2* (*TE2*) includes any device that isn't ISDN ready. This category includes the equipment you now use for analog communications. Any device in this class, such as a modem, requires an adapter to work with ISDN. A growing number of ISDN equipment vendors offer products that consolidate TE2 adapter and NT1 devices into a single unit.

Terminal Adapter (TA)

The functional device called a *Terminal Adapter (TA)* allows analog voice and data devices to work through an ISDN connection. The TA device is a protocol converter that adapts equipment that's not designed for ISDN. Over time, as equipment becomes ISDN ready (TE1), TAs will disappear. But for the foreseeable future, they're an important cornerstone of ISDN.

CPE vendors market terminal adapter devices that include the NT1 function as well as support other devices. For example, using a terminal adapter you can plug an analog telephone, a fax, and an ISDN adapter into your PC. This type of product controls the traffic from different devices sharing the same ISDN line and is called an NT1 Plus device.

ISDN Reference Points Primer

ISDN reference points define the communication between the different devices and the parameters for the functional devices. The four protocol reference points that are commonly defined for ISDN are called R, S, T, and U. Understanding ISDN reference points is important because most CPE vendors refer to their equipment in terms of the reference points it embodies.

The R Reference Point

The R reference point lies between the terminal equipment 2 (TE2) device and a terminal adapter. There are no specific standards for the R reference point, so the TA manufacturer determines and specifies how a TE2 and TA communicate with each other.

The S and T Reference Points

The S reference point lies between ISDN user equipment (TE1 or TE2 with a TA) and the NT1 device. The T reference point lies between customer site PBX switching equipment (NT2) and the local loop termination (NT1). In the absence of the NT2 — which is the case for BRI users — the user-network reference point is usually the S/T reference point. The S/T reference point is the one of two reference points that most ISDN equipment vendors incorporate in their devices. An S/T device requires a stand-alone NT1 device to work with your ISDN connection.

The U Reference Point

The U reference point is where the local telephone company network arrives at your doorstep up to the NT1 device. The U-interface is also called the *U-Loop* because it represents the loop between your premises and the telephone company's CO. ISDN devices made for the U-interface include a built-in NT1 function.

Where to Now?

Learning about ISDN is like peeling away the layers of an onion. At times, like working with an onion, it can make you cry. Wipe your eyes and let's move to the next layer. The next chapter explains the nuts and bolts of establishing an ISDN connection through your local telephone company.

Chapter 3

Making the ISDN Connection

In This Chapter

- Get started with an ISDN connection plan
- Check ISDN availability and your telephone company switch type
- Determine your ISDN service needs based upon your customer premises equipment (CPE)
- Figure out your ISDN service costs
- Order your ISDN service

In the olden days of analog communications, getting started was simple. You ordered a POTS line, plugged in your equipment, then started communicating. Establishing ISDN service is a more complex undertaking. It involves making a bunch of decisions *before* you actually establish the service. This chapter lays out the big picture of what it takes to get connected to ISDN as a foundation for getting into the specifics we'll discuss later in the book.

Getting Started with an ISDN Connection Plan

The process of establishing an ISDN connection involves several components that converge to define your actual ISDN service. Your journey toward digital enlightenment via ISDN involves the following steps.

- Finding out if ISDN service is available at your location and what type of switch the telephone company uses at the local CO.
- Determining your ISDN service needs, including what equipment and related BRI configuration options you plan to use.
- Figuring out what ISDN service will cost.
- Ordering your ISDN service from the telephone company.

Checking ISDN Availability and Switches

The first step involves checking the availability of ISDN service in your area. Without local service, the remaining steps are moot unless you plan to relocate. If ISDN service is available in your area, you need to find out the type of switch used by the telephone company for the ISDN line. You need to know the type of switch to determine how many devices you can use as well as to determine other configuration settings for your ISDN equipment.

Checking ISDN Availability

The absolute first step in getting an ISDN connection is determining whether it's available in your area. Telephone companies differ considerably in their implementation of ISDN service. Figure 3-1 shows the projected ISDN deployment for RBOCs for 1995 and 1996 based on the percentage of POTS lines that support ISDN. You can use this chart to gauge the responsiveness of a telephone company to ISDN service, though it's only a rough gauge. Ultimately, you need to get down to specifics of whether ISDN is available to your premises by contacting the telephone company.

Some RBOCs offer a service called *virtual ISDN*. This lets a telephone company offer ISDN service from a CO that doesn't have the switches to handle ISDN service. The telephone company routes the local loop termination at your local CO to another CO via an internal high-speed data link.

Determining the Switch Type

The switch at the telephone company's CO defines the capabilities of your ISDN service. The switch type determines the number of ISDN devices you can use on the line and other features. It's essential to determine in advance the type of switch the telephone company will use for your ISDN service.

The three main types of ISDN switch software options for most telephone companies are: AT&T 5ESS Custom, AT&T 5ESS NI-1 (National ISDN-1), and NT DMS-100. Before you determine the applications you plan to use on your ISDN line, you must know what type of switch the telephone company will provide. Different switches have different capabilities for handling multiple ISDN devices and other configuration options. Table 3-1 lists the number of devices supported by each type of switch.

You also need to know the switch type to configure your CPE to work with your ISDN line. Most CPE documentation includes specific configuration information for each of these switches.

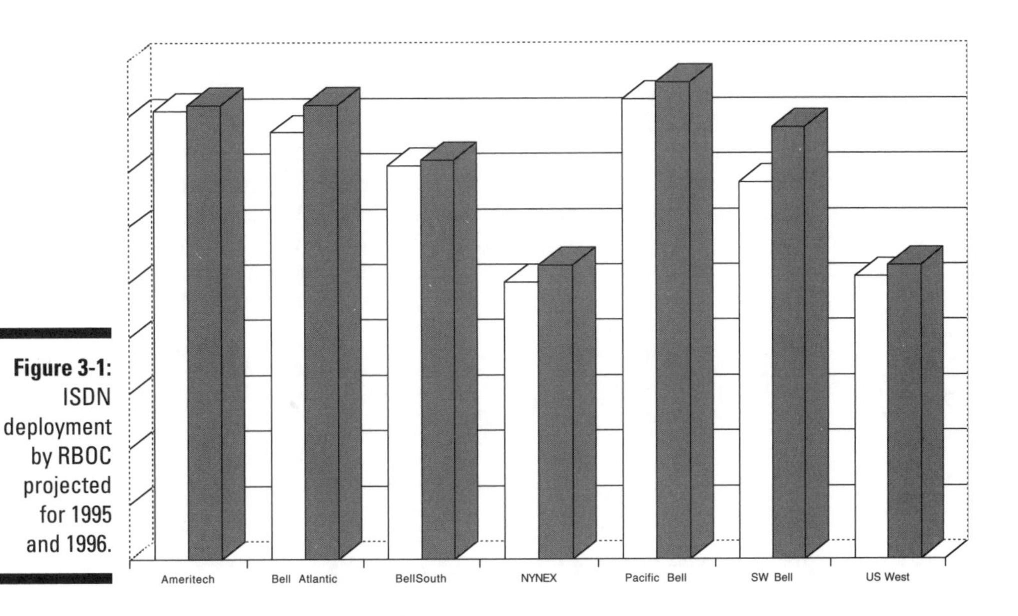

Figure 3-1: ISDN deployment by RBOC projected for 1995 and 1996.

Table 3-1: The number of devices supported by each type of telephone company switch

Switch type	*Number of ISDN devices supported*
AT&T 5ESS NI-1	eight devices
AT&T Custom	eight devices
NT DMS-100	two devices

All switches are not created equal. If you have a choice, ask for the AT&T 5ESS Custom. This switch offers the most flexibility in configuration options.

Getting Answers from the Telephone Company

Getting information about ISDN service can be a hit-or-miss proposition depending on your local telephone company. A few companies offer ISDN service centers staffed with knowledgeable people. Others make getting answers about ISDN more difficult. Table 3-2 lists telephone numbers for getting ISDN information from the RBOCs and ITCs.

Most telephone companies offer at least an automated voice system for checking ISDN availability. The automated system lets you check by using your Touch Tone telephone to enter your area code and exchange number (the three numbers after the area code).

One way or another, you need to check with your local telephone company to get a confirmation that ISDN service is available to your premises. Some companies have this information available in databases and can tell you right away. Others must perform a loop qualification, which can take days to complete. A *loop qualification* is the process through which the telephone company actually checks your local loop to see if it supports ISDN.

Don't trust automated systems for your final answer about ISDN availability. Even if the system says ISDN is available in your area, it may not be available at your premises because of the distance from the CO.

A Tale of Two Telcos

The quality of service in the delivery of ISDN can vary widely from one telephone company to another. My personal experience provides a real-world example.

Before moving to Boston, Massachusetts, I lived in Menlo Park, California. Pacific Bell is the telephone company in California and NYNEX is the telephone company for New England. PacBell is one of the most aggressive of the RBOCs in providing ISDN connections at affordable prices. PacBell provide a toll-free ISDN service center for establishing ISDN service. I got an ISDN connection to my home at a low residential rate of about $15 per month plus an installation fee of $195. ($125 was a refundable deposit if I kept the ISDN service for two years.)

I moved to the Boston area and decided to get an ISDN line from NYNEX for my new residence. What I experienced was not digital enlightenment, it was a digital nightmare. NYNEX offered an automated system to tell people what areas were served by ISDN. From the point of determining availability, the service dropped off. After a series of frustrated calls to ISDN product management and others at NYNEX, I finally obtained an ISDN line eight weeks after my first inquiry. Not only was getting the service difficult, the cost was painful. It cost almost $500 to install the line. In addition, NYNEX doesn't offer a residential rate, so it costs more than $40 per month.

The moral of this story is that different telephone companies have different attitudes about ISDN service, and it shows. In the case of NYNEX, it appears that top management doesn't support ISDN. As a consumer, be prepared for some telephone companies like NYNEX to be downright unhelpful. But don't give up.

Table 3-2: Where to get ISDN information from RBOCs and ITCs

Telephone Company	*Call*
Ameritech	800-832-6328 to access a menu-driven, 24-hour voice/fax-back system.
Bell Atlantic	800-570-4736 for the ISDN Sales and Tech Center; 800-843-2255 for small businesses.
BellSouth	800-428-4736
Cincinnati Bell	513-566-3282
GTE	800-888-8799 to access a 24-hour, menu-driven voice/fax-back system; 214-718-5608 by voice.
Nevada Bell	702-334-4811 for small businesses; 702-688-7100 for large businesses.
NYNEX New York	800-438-4736 to access a 24-hour, menu-driven voice/fax back system; 914-644-5152 by voice.
NYNEX New England	800-438-4736 to access a 24-hour, menu-driven voice/fax-back system; 617-743-4513 by voice.
Pacific Bell	800-472-4736 for the ISDN Service Center; 800-995-0345 to access a menu-driven, 24-hour voice/fax-back system.
Rochester Telephone	716-777-1234
Southwestern Bell	800-792-4736, Austin; 214-268-1403, Dallas; 713-567-4300, South Houston; 713-537-3930, North Houston; 800-992-4736, all other locations.
US West	800-246-5226

Additional resources for getting ISDN availability information are provided in Chapter 23. ■

Determining Your ISDN Service Needs

Once you know about ISDN availability and the type of switch your telephone company uses for ISDN, you need to determine your ISDN service needs. This process is a function of the ISDN applications and equipment you plan for your ISDN connection. The term used for determining the configuration of your BRI line is commonly referred to as provisioning your line.

Planning Your CPE Needs

Each application and its associated equipment require special configurations that you must tell the telephone company about when ordering ISDN service. The telephone company uses this information to program its switch at the CO to work with your equipment. The more applications you plan to use with your ISDN connection, the more configuration specifications you need for the ISDN line. Because you can — and in most cases will — use multiple ISDN devices on your ISDN connection, you need to coordinate your CPE activities into a unified ISDN system. Chapter 4 deals with the specific planning issues for developing a CPE strategy to make the best use of your ISDN line.

ISDN Ordering Codes

To simplify the process of ordering and configuring your ISDN connection, a system called *ISDN Ordering Codes* streamlines the process of matching your ISDN service to your equipment. An ISDN Ordering Code represents a pre-established set of network services associated with specific CPE in support of a particular application. An example of ISDN Ordering Codes is the instruction sheets included with the Intel ProShare personal video conferencing system. This template is called Intel Blue. You simply tell this to the telephone company staff and they know exactly how to configure the line for you.

To order ISDN service with the telephone company using ISDN Ordering Codes, you need the unique code name supplied by the CPE vendor. This code identifies the network services that are compatible with the CPE. Over 40 CPE vendors now support the concept and usage of ISDN Ordering Codes, and more than 100 codes have been registered. Currently, more than 60 ISDN products have ISDN Ordering Codes, and 10 Service Providers have committed to supporting them.

The ISDN Ordering Codes system was developed by the National ISDN User Forum and the Corporation for Open Systems. Bellcore has taken over the responsibility for maintaining the ISDN Ordering Codes system to facilitate a centralized source for vendors and telephone companies to provide and distribute ordering code information.

Choosing Your Channel Configuration

The Basic Rate Interface standard is two B channels and one D channel. However, different telephone companies offer a menu of BRI channel configurations. Table 3-3 shows the wide variety of available BRI configuration options that NYNEX offers.

Your BRI channel configuration determines the type of information that gets transmitted through each B channel. You determine the channel configurations when you order your ISDN service. If you make any changes, your telephone company will charge a fee. The following are the available channel configuration options.

Table 3-3: BRI configuration options from NYNEX for Massachusetts

Interface Type	*Interface Configuration*
0B+D	D Channel Only
1B	1B Voice
1B	1B Data
1B	1B Alternate Voice/Data
1B	1B Packet Data
1B+D	1B Voice D Packet Data
1B+D	1B Data D Packet Data
1B+D	1B Alternate Voice/Data D Packet Data
2B	1B Voice 1B Data
2B	1B Voice 1B Packet
2B	2B Data
2B	1B Data 1B Voice/Data
2B	1B Data 1B Packet Data
2B	1B Voice/Data 1B Packet Data
2B+D	1B Voice 1B Packet Data
2B+D	2B Data D Packet Data
2B+D	1B Data 1B Voice/Data D Packet Data

- *Circuit switched voice (CSV) only.* This option allows only voice traffic for the B channel. This is a restrictive option that doesn't allow you to get the full bandwidth of the BRI connection because it ties up a B channel for just voice.
- *Circuit switched data (CSD) only.* This option allows circuit data only for data transmission speeds up to 64 Kbps (uncompressed) for each B channel.
- *Alternate circuit switched voice/circuit switched data (CSV/CSD).* This option allows either circuit data or voice communications to be carried over the channel. This channel configuration allows for the most versatile use of an ISDN line.
- *Packet data only.* This option allows for only X.25 packet data.

In most cases you'll want to order the basic 2B+D channel configuration, with one B channel as alternate voice/data, and the other as data only or alternate voice/data. If you don't need 16 Kbps of X.25 capacity for the D channel, request a 2B fconfiguration.

Every BRI connection has a D channel for out-of-band signaling. Beyond the requirement of signaling, the D channel offers 16 Kbps of X.25 packet data capacity. Some telephone companies let you specify whether you want the extra X.25 capacity on the D channel. If you specify not to have the X.25 capacity, your BRI line still uses a D channel for out-of-band signaling, but you don't have the 16 Kbps of X.25 packet data available.

Point-to-Point or Multipoint?

An ISDN line can be either point-to-point or multipoint. Point-to-point configuration refers to the operation of one device on an ISDN line. Multipoint configuration refers to the operation of multiple devices on an ISDN line. In most cases you'll be working with multiple ISDN devices, so the standard configuration is multipoint.

Within a multipoint configuration, each switch allows a certain number of devices to be connected to the line. An important criterion for ISDN equipment is to ensure that it supports the multipoint protocol. If an ISDN device only supports point-to-point, you won't be able to use the line for other applications.

Call Appearances

If you're using an ISDN telephone or analog telephone on your ISDN telephone line, you'll want to order call appearances. Call appearances can be separate telephone numbers or multiple appearances of the same telephone number. They allow for an increased volume of telephone calls or for incorporating call management features such as call forwarding and three-way calling at your site.

Call appearances can act like POTS call waiting, but with more capacity for multiple calls on a single ISDN line. For example, a call arrives while you're talking to someone, you tap the flash button and talk to the person on the other line. Another call arrives and you put the second person on hold or route it elsewhere, such as to an answering machine or another telephone.

Different telephone companies charge different rates for call appearances. NYNEX charges a one-time installation fee of $30 for 10 call appearances plus $5.20 per month. You can order additional call appearances in multiples of 10.

Figuring Your ISDN Service Costs

Let's get down to the bottom line: What's all this digital enlightenment going to cost? The total cost of your BRI line depends on a variety of fixed and variable costs. Many of these costs are similar in structure to those of a POTS connection. ISDN service costs includes a one-time installation charge, recurring monthly charges, and usage charges. To help you analyze how much an ISDN connection will cost, the following sections explain each of the factors that make up the cost of your total ISDN service package.

See Appendix A for a summary of specific ISDN tariffs for RBOCs.

ISDN Economics 101

One thing is certain about ISDN service: It costs more than POTS, but it also delivers more capabilities. The best way to get the most for your money with ISDN is to use it for multiple communication tasks. ISDN service lets you perform both analog and digital communications on the same line, so you may be able to eliminate some of your POTS lines. The enhanced call management features of ISDN let you also handle multiple voice communications over a single line.

ISDN saves you connection time and costs. The expanded bandwidth of ISDN lets you transfer more data in less time. For example, if you transfer many large files, the cost performance of ISDN over POTS and a modem is substantial. For work in cyberspace, ISDN can substantially cut connection time, saving both telephone call charges and charges levied by online and Internet service providers.

Business vs. Residential Rate

As is the case for analog telephone service, ISDN service can be based on a business or residential tariff. Generally, the cost of business telephone service is higher than residential service. Some RBOCs, such as Pacific Bell, offer residential ISDN service at cheaper rates than business tariffs. Other RBOCs charge a business tariff for ISDN whether it's used for a business or a residence. Not only is the residential rate usually cheaper for installing an ISDN line, recurring and usage charges are also cheaper in many cases.

Installation Charges

The installation of an ISDN line averages about $150. Your exact installation charge depends on your channel configurations, whether you're installing a new line or converting an existing POTS line for ISDN, and any wiring on your premises. Table 3-4 shows average BRI installation charges for the RBOCs. Some telephone companies, such as PacBell, waive the installation charge if you continue ISDN service for two years. For many telephone companies, if you cancel your ISDN service, you'll be charged a termination fee that can run over $100.

If you have the telephone company wire your home or business — that's twisted-pair wiring — they charge around $55 to $75 per hour. You can also use independent contractors or do the wiring yourself within your own premises for ISDN, as explained in Chapter 4.

The U-interface, or the line the telephone company terminates at the demarcation, is a two-wire interface, so it uses an RJ-11 jack and cord. This is the standard cabling used by analog telephones. The S/T-interface, on the user's side of the NT1, uses the RJ-45 — an eight-pin jack. Most CPE vendors provide a short RJ-45 cord to go from the NT-1 to the equipment. You can plug an RJ-11 connector into an RJ-45 jack.

Don't let the telephone company installer sell you an RJ-45 jack for ISDN service — you don't need it. All you need is the standard RJ-11 jack, which you can probably install yourself. The telephone company will also charge you a lot more for the RJ-45 jack than that for which you can buy it. For example, NYNEX charges $12.50 for the RJ-45 jack. See Chapter 4 for more information on ISDN wiring. ■

Recurring Monthly Costs

ISDN service has a recurring monthly charge. The amount of this fee depends on your BRI configuration. Table 3-5 shows the average monthly recurring costs charged by RBOCs for a BRI connection. These charges don't include usage.

Table 3-4: Average BRI installation cost

RBOC	*Average BRI installation cost*
Ameritech	$158
Bell Atlantic	$181
BellSouth	$205
NYNEX	$164
PacBell	$195
SWBell	$154
USWest	$85

Table 3-5: Average monthly recurring costs for a BRI connection

RBOC	*Average monthly BRI cost*
Ameritech	$52
Bell Atlantic	$47
BellSouth	$38
NYNEX	$52
PacBell	$27
SWBell	$56
USWest	$67

Usage Charges

Beyond the installation and monthly costs of ISDN service, you will have usage charges. Like their POTS counterparts, these charges can easily dwarf your recurring monthly costs, depending where you're calling. The local telephone company charges you for local calls and the long distance telephone company bills you for long distance calls.

Unlike POTS service in many parts of the U.S., your local and long distance circuit switched ISDN usage charges don't arrive in a consolidated bill. You'll receive a separate long distance bill from the long distance carrier that you specified at the time you ordered ISDN service. However, long distance voice charges may appear in a long distance section of your local telephone bill.

ISDN toll charges generally follow the formula used for POTS service that breaks down the charges according to the time the call is made and the distance of the call. The three standard calling time categories include Day, Evening, and Night. The Day rate is the most expensive and the Night rate is the cheapest. The toll rates for ISDN are just as complex as for POTS, and ISDN brings some new charges. For example, if you're using both B channels during a connection, you'll be charged twice the published tariff because the two B channels are treated as two separate lines. If you're using the D channel (or a B channel) to send X.25 data, the telephone company charges various rates based on the amount of data you move through the channel.

Charges for switched data are usually handled differently from those for voice. For voice, standard POTS voice rates for residential or business service prevail in most cases. For circuit switched data usage, many RBOCs charge a different tariff. Table 3-6 shows how NYNEX charges for circuit switched data. Keep in mind that if you're using both B channels in an ISDN call, the rate is double.

Table 3-6: Usage Charges for NYNEX New England

Calling Area	*Charge Per Call*	*Per Minute Charges (or fraction)*
Within Calling Area		
Zone 1 Exchanges	$.0603	$.016
Zone 2 Exchanges	$.0603	$.016
Outside Calling Area		
Day Rate	$.01	$.105
Evening Rate	$.01	$.055
Night Rate	$.01	$.036

Beyond the local usage charges are long distance charges. Table 3-7 lists the toll charges for AT&T long distance ISDN service for circuit switched data, which is based on a mileage formula. These rates reflect the charge per minute for each B channel; so again, if you're using both B channels, the charge is double.

Ordering Your ISDN Connection

The final part of establishing ISDN service is contacting the local telephone company to order your service. You should order your ISDN service only after you've assembled the information on the devices you plan to use, the BRI channel configurations you want, and any other specifications.

Table 3-7: AT&T's Day Rate for each B channel

Mileage	First Minute	Additional Minute
0-55	$.2790	$.1990
56-124	$.3090	$.2290
125-292	$.3310	$.2510
293-430	$.3530	$.2730
431-925	$.3790	$.2990
926-1910	$.3880	$.3080
1911-3000	$.4000	$.3200
3001+	$1.1470	$1.0670

As you'll recall, ordering an ISDN connection involves exchanging information with the telephone company. This information includes line and switch configuration information for each CPE you plan to use on the line. From the telephone company you receive information for configuring your CPE to work with the ISDN connection. The following sections explain what information you'll receive from the telephone company after you give them your configuration information. Be aware that the telephone company may take some time to get your line established and give you back any information you need to configure your devices.

SPID on Your ISDN Connection

SPID stands for *Service Profile Identifier,* a number assigned by your telephone company to each device connected to the ISDN line. SPIDs let the telephone company switch know which ISDN services a given device can access. SPIDs are important because you need them to set up your ISDN equipment to work with the ISDN line. A unique SPID is required for every device on a multipoint ISDN line or one for each B channel. An example of two B channel SPIDs might be as follows: For the first B channel the SPID might be 50862890210000 and for the second B channel the SPID might be 50862809620000.

Typically, a SPID is a telephone number with several digits after it. When you configure your software for an ISDN device, you usually need to configure the software to tell it the SPID assigned to it. Multipoint ISDN lines can handle multiple SPIDs. The AT&T 5ESS Custom switch supports up to eight SPIDs, as does the AT&T 5ESS National ISDN switch. The NT DMS-100 National ISDN 1 supports only two SPIDs.

Directory Numbers

A *directory number (DN)* is the address or telephone number for the ISDN line assigned by the telephone company. Each ISDN line receives one telephone number, called the *Primary Directory Number (PDN)*. If the CO switch is an AT&T 5ESS Custom, you only need one DN per device. If the CO switch is an AT&T 5ESS NI-1 or NT DMS-100 NI-1, you need a separate DN for each of the two B channels.

Where to Now?

The next phase in your path to digital enlightenment is developing a CPE game plan to get the most from your ISDN service. To do so, you need to understand how different devices work — or don't work — together. The next chapter explains the key pieces of an ISDN connection at your premises, including ISDN wiring, CPE hardware options, and how to develop your CPE game plan.

Chapter 4

Developing Your CPE Game Plan

In This Chapter

- Wiring for ISDN
- Equipping yourself for ISDN
- Developing your CPE strategy

On your side of the ISDN connection is a cornucopia of consumer premises equipment (CPE) options. Before you go off to explore specific applications and related CPE, you need to understand how different CPE pieces interact with each other. Grasping how to manage this integration is essential to developing an integrated ISDN solution. This chapter explains how to get wired for ISDN, the main CPE categories, and key factors for developing your CPE game plan.

Getting Wired for ISDN

ISDN uses the same wiring as that used for POTS, but wiring configurations for ISDN operate differently. You need to understand several important wiring issues before dealing with your ISDN CPE. The following sections explain ISDN wiring issues related to laying your cable foundation for implementing your CPE plan.

Converting an Existing Line or Getting a New Line

One of the first ISDN wiring issues is whether you want to convert an existing POTS line to ISDN service or to add a completely new line. Depending on whether your premises is a business site or a residence, the location can impact which option you can use.

In most cases, business sites don't have restrictions for bringing in additional new lines. For residences located outside metropolitan areas, the number of telephone lines available may be restricted to two lines. The standard single four-wire cable allows for two lines. One line uses the red and green wires, while the other line uses the yellow and black wires. Most homes in metropolitan areas can add additional lines beyond two. Check with your telephone company to see if you can bring in an additional line before deciding whether to convert an existing analog line.

If you can't bring another line into your home, you'll need to convert an existing POTS line to ISDN. Don't worry, you can use an ISDN line for analog communications. For example, if you use a second analog line for telephone, fax, or modem communications, you can still use these devices with an ISDN line for high-speed digital communications.

In most cases you don't want to convert you primary analog voice line to ISDN service. If the telephone company can add a new line to your premises, you may want to do it. It takes time to climb the ISDN learning curve, so you may not want to hold your primary analog communications line hostage as you make the transition to ISDN. Additionally, you may want to use your ISDN line exclusively for data transmission and not tie it up with incoming voice calls.

Wiring from the Telephone Company to Your Premises

The U reference point covers the wiring from the telephone company's CO to your premises as well as your site's internal wiring up to the NT1. In the United States, the division of telephone wiring responsibility between the telephone company and the end user is different from the U-interface. The telephone company is responsible for any wiring to your doorstep, but from that point it's your responsibility.

The *demarcation point* is the dividing line between the telephone company's wiring and the premises wiring. The physical device that provides the means to connect the telephone company's wire to the premises wire is the *Network Interface* (*NI*). The demarcation point and Network Interface box are the same for POTS and ISDN service. However, if your premises are in an older structure, you may not have a Network Interface box. As a result, the incoming telephone line connects directly to a device called a *Protector Block*.

The FCC places a restriction commonly referred to as the *12-inch rule* on both the customer and the telephone company. Basically, this rules states that the Network Interface box must be located within 12 inches of the Protector Block. If your house doesn't have a Network Interface box, you must have the telephone company install one. The FCC prohibits customers from working at the Protector Block. Only the telephone company can install the Network Interface device or any device directly wired to the telephone line.

Wiring Your Premises

From the demarcation point, your telephone wiring is your responsibility. You can have the telephone company or an independent contractor do your inside wiring for a new line, or you can do it yourself. If you're using one of two pairs of wires in a line, you need to know which pair of wires is used for which line.

If you have a choice, bring in your ISDN line as a separate line and use just one of the wires for the ISDN line. It makes it easier to install and set up ISDN equipment. ■

If your existing wiring has worked properly for analog service, it should continue to work for ISDN service. You may run into problems if the existing wiring is complex, such as office wiring originally installed for older key-based telephone systems. If this is the case, you'll need to consult with your telephone company or wiring contractor.

The EIA/TIA (Electronic Industries Associations and the Telecommunications Industry Association) standard for wiring analog and ISDN service requires an *unshielded twisted-pair* (*UTP*) cable of category 3 or above — 24-gauge for new residential wiring. The designation unshielded refers to the fact that the sheath does not include an electrical shield.

Making Connections

At the end of your wiring are the connectors. Standard analog wiring uses RJ-11 modular connectors that snap into a surface-mounted or flush-mounted jack. ISDN uses the same RJ-11 connectors and cabling. ISDN devices use RJ-45 connectors to connect to the NT1 device. The RJ-45 is an eight-position jack commonly used in business telephone systems, but usually not used in residential wiring. The NT1 device typically has an RJ-45 jack, which you can plug into the RJ-11 cable from your inside wiring.

Configuring Your ISDN Wiring

For POTS service, you normally add a new telephone or other analog device by simply splitting the RJ-11 jack or tapping into the existing line and adding a jack. This type of wiring arrangement is called a *daisy chain*, as shown in Figure 4-1. While daisy-chain wiring works for POTS, it doesn't work for an ISDN line.

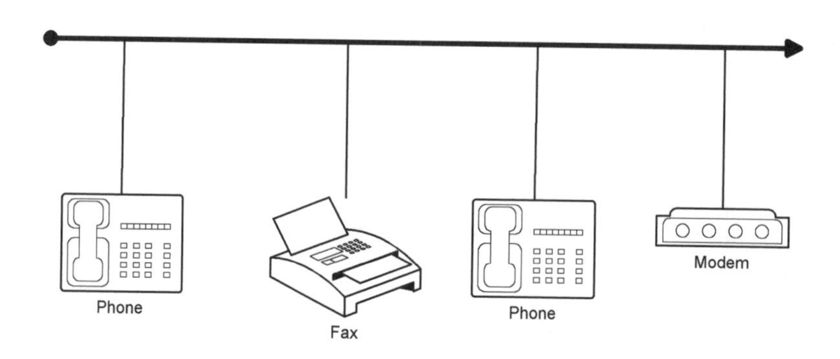

Figure 4-1: The typical POTS wiring scheme allows you to tap into an existing line at any point to add another telephone or other analog device.

Your ISDN service comes in as a single line and terminates at the Network Termination device (NT1). You can't add any ISDN devices to the line before it gets to the NT1. Once the ISDN line terminates at the NT1, you can add multiple ISDN devices using multiple RJ-45 connectors. The wiring arrangement for ISDN is the *star configuration* in which the wiring to any individual piece of terminal equipment comes from a central point, which is the NT1 or NT1 Plus device, and goes to the terminal equipment. Figure 4-2 shows this type of wiring configuration.

The normal wiring scheme for connecting ISDN devices to the NT1 device via RJ-45 wiring is called a *passive bus* configuration. Most small businesses and individuals can use this configuration. The passive bus arrangement allows the connection of multiple S/T-interface or analog devices to share an ISDN line without repeaters to boost the digital signal. Chapter 5 explains the specifics of wiring ISDN devices to your NT1 device.

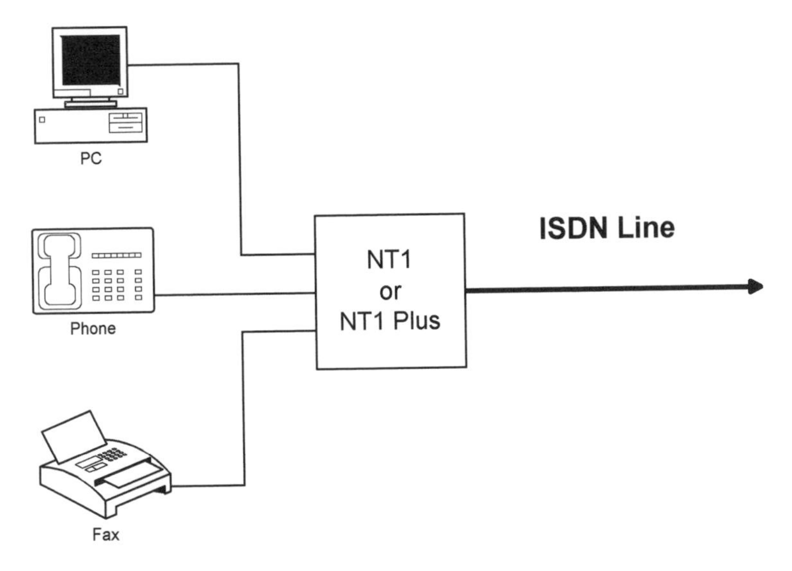

Figure 4-2: The wiring for ISDN comes from the user's ISDN devices to a central network termination device.

Equipping Yourself for ISDN

To work with ISDN requires setting up a configuration of different devices. These devices cover the functions required for ISDN, such as network termination, as well as devices for specific ISDN applications, such as video conferencing. ISDN applications and equipment continue to evolve, but the following CPE categories remain fairly stable.

- NT1 device that provides the network termination and ISDN line-powering functions.
- NT1 Plus device that provides the network termination and ISDN line powering function as well as terminal adapter function.
- ISDN telephones for voice communications.
- Remote access adapter cards or stand-alone devices for connecting your PC to an office LAN or any network via ISDN.
- Video conferencing systems that include a video capture card, ISDN adapter, and video camera.

The following section presents an overview of the main types of devices and applications you'll typically use on your ISDN in one configuration or another.

Stand-Alone NT1 Devices

A stand-alone NT1 provides just the NT1 and line-powering functions for your ISDN connection. It allows you to connect multiple ISDN S/T-devices to your ISDN line. The NT1 device is strictly a conduit for connecting the U-interface to your ISDN devices via RJ-45 connectors. A stand-alone NT1 device is useful if you don't need to connect any analog equipment to your ISDN line.

Chapters 5 and 6 explain NT1 devices in detail. Chapter 18 provides information on leading NT1 vendors. ■

NT1 Plus Devices

An *NT1 Plus* device is often the best starting point for working with ISDN if you plan to use analog devices, such as a telephone, fax, or modem. These devices are hybrid devices that include the NT1 and terminal adapter functions. As you recall, the terminal adapter (TA) functional device allows non-ISDN devices to communicate via ISDN. The configuration of the ports available depends on the product. Because NT1 devices let you work with ISDN's voice services, they require more effort to set up.

Chapters 5, 7, and 8 explain working with NT1 Plus devices. Chapter 18 provides information on NT1 Plus vendors. ■

ISDN Telephones

ISDN telephones are voice devices designed specifically to take advantage of ISDN's powerful call management features. These features let you perform a multitude of calling tasks and include displaying Caller IDs, handling multiple incoming calls from a single line, transferring calls to other numbers, establishing conference calls, and more. Using a system of ISDN telephones, a small business can create a sophisticated call management system from desktop telephones that before ISDN was only available to large organizations that could afford expensive PBX equipment. A typical digital telephone with a good collection of built-in call management support now sells for under $250.

Chapter 5 explains working with ISDN's voice communications features and ISDN telephones. Chapter 18 provides information on leading ISDN telephone vendors. ■

Remote Access Devices

Remote access devices allow your PC to connect to another PC, LAN, or online service via ISDN. They come in two flavors: a bus adapter card and a stand-alone unit. Using a remote access device, you can connect to the Internet via any service provider that supports ISDN connections, or you can connect to your office LAN to work remotely. Part III explains connecting your PC to cyberspace via ISDN.

ISDN bus adapter cards are the least expensive remote access solution. Bus cards use the PC bus configuration to communicate from your PC to the ISDN adapter. Adapter cards support either Ethernet or serial communication. These cards fit into an expansion slot in your PC and are available as U- or S/T-interface devices. Bus adapter cards, unlike bridge devices, don't require a network adapter card. Many of these stand-alone cards include analog ports for voice and modem communications.

A stand-alone ISDN bridge looks like a standard modem and requires a LAN adapter in the PC. A *bridge* connects separate physical networks into a single logical network that behaves as though it were a single physical network. In the case of ISDN bridges, the PC communicates via Ethernet to ISDN. The Ethernet card connects to the ISDN bridge via thin Ethernet coaxial cable or an RJ-45 cable. At the low end of the bridge solutions, are bridges that connect only one PC, Macintosh, or workstation to ISDN. At the high end of the market are bridges and routers that connect LANs to LANs. LAN-to-LAN connections via ISDN are beyond the scope of this book.

Part III covers remote access options for telecommuting or surfing the Internet via ISDN. Chapters 21 lists Internet service providers offering ISDN connections, while Chapter 22 provides an extensive list of remote access products and vendors. ■

Video Conferencing Systems

Desktop video conferencing systems are one of the most exciting uses of ISDN. The bandwidth that ISDN delivers makes real-time video communications come alive. Video conferencing systems typically include a video compression card, a sound card, a video camera, and software. The cards plug into a PC and provide both video and audio services. Because most PC video conferencing systems comply with standards, these systems allow you to communicate with other people using different systems. These PC-based systems can also communicate with existing room-size systems that are in wide use today.

Part IV explains working with desktop video conferencing systems. Chapter 20 list leading video conferencing vendors and other resources. ■

Developing Your Basic CPE Game Plan

An ISDN CPE game plan is essential to get the most from your ISDN service. Understanding a few core CPE fundamentals provides you with the foundation for choosing and integrating CPE into a complete system. The following sections present key CPE strategy issues to keep in mind as you assemble the pieces of your ISDN connection.

All Digital Roads Lead to NT1

You can have only one NT1 function to terminate your BRI connection. The placement of the NT1 device at your premises is the key factor in determining the number of applications you can use on an ISDN line. As a strategy for getting the most from an ISDN line, you want to be able to work with multiple devices. The optimal device for getting the most from your ISDN connection is using either a stand-alone NT1 device or an NT1 Plus device.

On the surface, getting NT1 built into a device like an ISDN PC adapter card designed for connecting to a LAN or the Internet seems convenient. However, using an ISDN device that includes the built-in NT1 and offers no ports for other devices limits your ISDN connection to the application of that one device.

Using an NT1 Plus device that includes the built-in NT1 function with ports for connecting analog and ISDN equipment via an S/T-interface port allows you to make full use of your ISDN service. A *port* is a location for passing data in and out of a device.

As a strategy for getting the most from an ISDN line, use a stand-alone NT1 device or NT1 Plus device instead of getting an NT1 built into a single application device, such as a remote access adapter card. ■

U- or S/T-Interface

ISDN equipment comes in two flavors, U and S/T; CPE vendors typically sell their products in both flavors. ISDN equipment made for the U reference point means the NT1 functional device is built in. ISDN equipment made for the S/T-interface requires the NT1 function to connect to ISDN. The significance of these two reference points is pivotal to your CPE game plan. The bottom line is that you can't use multiple U-interface devices on the same ISDN line, because it includes the NT1 function. On the other hand, you can use multiple S/T-interface devices on an ISDN line, because they don't include built-in NT1 devices.

In most cases you should purchase only S/T-interface ISDN equipment that connects to a stand-alone NT1 or NT1 Plus device.

Interoperability Awareness

Interoperability remains a leading problem in the full implementation of ISDN. Interoperability means that related devices from different vendors can work together. In the analog communications world, using a modem from one vendor to connect to a modem from another works because of standards that all modem manufacturers follow.

Some ISDN applications and their related equipment lack universal standards for vendors. The good news is that this situation is rapidly changing as standards for applications are put in place. For example, video conferencing products based on the H.320 standard are now common. This allows you to use a Vivo320 system to connect to someone using AT&T's Vistium Personal Video System.

In the area of ISDN PC adapters, the Point-to-Point/Multipoint Protocol standard allows you to connect to the Internet (or any other TCP/IP network) and use both B channels.

Early ISDN telephones only worked with certain digital switches. For example, AT&T ISDN telephones only worked with AT&T 5ESS switches. But as the ISDN network opens up, digital telephone vendors are offering telephones that work with any telephone switch.

BONDING Power (No, It's Not James Bond)

BONDING stands for Bandwidth On Demand Interoperability Group, which is a method for combining multiple channels into a single channel. For ISDN, BONDING allows the two B channels to be combined into a 128 Kbps transmission for the same application. Combining multiple B channels allows more data to flow through an ISDN connection. Applications such as file transfers and video conferencing depend on BONDING. BONDING is embodied at the CPE level, meaning equipment vendors include it or don't include it as a feature of their products.

ISDN CPE Recipes

Understanding the pieces is one thing; understanding how to assemble them into a CPE package to use your ISDN connection to its fullest is another. Assembling all the pieces for an effective ISDN connection requires some visualization to see the layout options as defined by different applications.

To guide you through the rest of this book and visualize the possibilities for an ISDN system, the following sections present typical ISDN configuration scenarios. All these recipes take into account that you'll want to use ISDN for multiple services to justify its added expense. While you don't have to buy all the equipment at once, you need to build the right foundation that will allow you to add devices.

The Optimal PC Configuration for ISDN

ISDN is a powerful communications connection that requires matching PC power. Using a slow processor, an older graphics adapter, or too little memory causes bottlenecks that will affect the overall performance of your system. For example, downloading graphics to your systems at eight times the speed of a 14.4 Kbps modem is diminished if your graphics card is slow at redrawing your screen. Here is an optimal PC configuration for working with ISDN across all its available applications. In most cases, you can get by with less, but you'll pay in terms of performance. Your PC should include:

- A 486DX2 CPU at 66MHz or faster; or better yet, a Pentium CPU at 60MHz or faster.
- An accelerated VESA bus or PCI bus graphics card that supports at least SVGA with 800x600 resolution and 256 colors.
- At least 8MB of RAM; 16MB is better.
- One to three empty 16-bit ISA bus slots, depending on your ISDN applications.
- Microsoft Windows 3.1 or 3.11, or Windows for Workgroups 3.11, with MS-DOS 5.0 or later.

Basic Voice and Remote Access Package

If you're a telecommuter, work at a small office, or work at home, typically there are two ISDN applications that form a core package. These applications include remote access and voice communications. Remote access let's you connect to a LAN, the Internet, or any other host computer. Remote access devices come in two flavors: adapter cards and stand-alone units. You can add an analog or ISDN telephone to your ISDN line so you can handle voice communications on the same line. The following are some possible device configuration options for a combination voice and remote access package.

- A remote access adapter card with an analog telephone or headset port that connects to an NT1 device.
- A remote access card or stand-alone device with a built-in S/T-interface port for an ISDN telephone that connect to an NT1 device. Figure 4-3 shows this scenario using a remote access adapter card.
- A remote access device and an analog telephone that plug into an NT1 Plus device with a POTS and S/T-interface port.
- A remote access adapter card with an S/T-interface ISDN telephone that plug into an NT1 device, as shown in Figure 4-4.

The remote access voice combination can work in different ways depending on how you're using your remote access. If you're using both B channels for surfing the Internet, you can't use your line for voice unless your data transmission is reduced to using only one B channel.

You can also use a POTS port on an NT1 Plus device to plug in a fax machine or modem instead of a telephone. Or if the NT1 Plus supports two POTS ports, you can plug in both a telephone and either a fax or a modem.

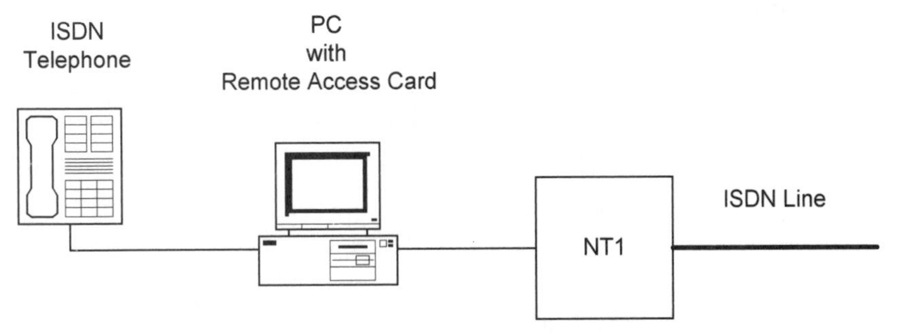

Figure 4-3: A remote access card with a built-in S/T-interface port for an ISDN telephone that connects to an NT1 device.

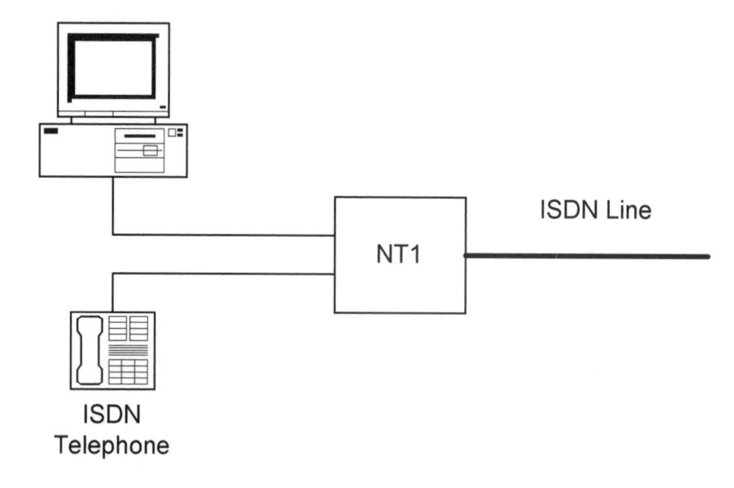

Figure 4-4: A remote access adapter card and an S/T-interface ISDN telephone that plug into an NT1 device.

Voice, Remote Access, and Video Conferencing

If you want to maximize your ISDN line, you can combine voice, remote access, and video conferencing. The following are some possible device configuration options for a combination voice, remote access, and video conferencing package.

- An S/T-interface remote access device and video conferencing system that plug into an NT1 Plus device with a POTS port for connecting an analog telephone (or any other analog device).
- A remote access card with an S/T-interface port, a video conferencing system, and an ISDN telephone that plug into an NT1 device as shown in Figure 4-5.

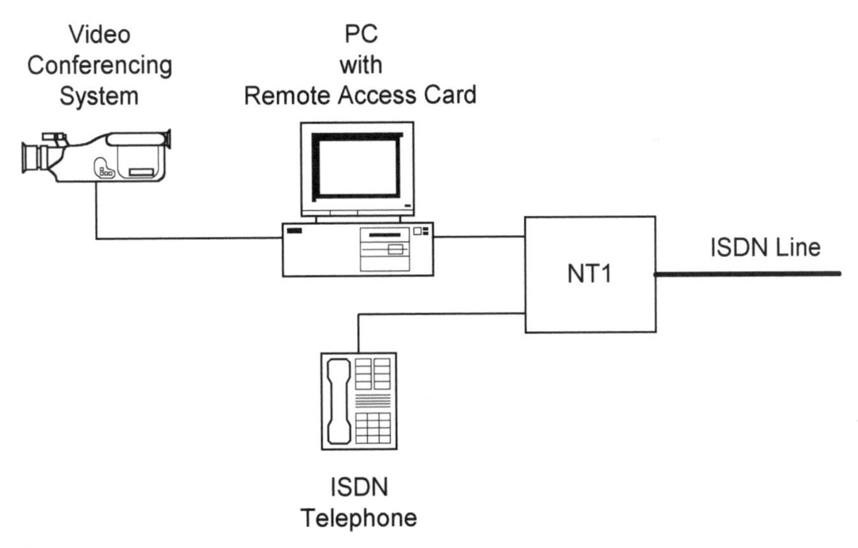

Figure 4-5: A remote access card with an S/T-interface port for plugging in a video conferencing system and an ISDN telephone together create a full-service ISDN connection.

And Away We Go . . .

Finally, there's light at the end of your ISDN fundamentals journey! Now it's on to something you can sink your teeth into — working with ISDN applications. The first stop on your application adventures is working with NT1 and NT1 Plus devices, and with ISDN telephones.

Part II

NT1s, NT1 Plus Devices, and ISDN Telephones

The 5th Wave

In This Part...

The first fork in the road to digital enlightenment involves choosing the type of network termination device that's best for you. This device defines your premises as a site on the global ISDN network. In the following chapters you'll learn what your network termination options are, how they work, and which one to use for what type of ISDN applications. You'll discover which device to use to connect ISDN-ready devices, or which network termination device to use to connect your existing analog devices to an ISDN line — equipment such as a telephone, fax machine, and modem. You'll experience working hands-on with both types of devices. Finally, you'll develop the skills to harness ISDN's powerful voice communications capabilities using either an ISDN or analog telephone.

Chapter 5

NT1, NT1 Plus, and ISDN Voice Communications Basics

In This Chapter

- Determining which way to go — NT1 or NT1 Plus
- Making digital calls with ISDN telephones
- Understanding ISDN call management features

The first step in getting started with ISDN involves deciding whether to use an NT1 or an NT1 Plus device. Both devices provide the network termination function for your BRI connection. The NT1 Plus adds terminal adapter functions so you can use analog or other devices with your ISDN connection. ISDN delivers powerful calling features to either an analog telephone connected to an NT1 Plus device or an ISDN telephone. This chapter covers NT1, NT1 Plus, and ISDN voice communications fundamentals.

Terminating Your ISDN Connection via NT1

ISDN is a giant network, with the NT1 device defining your BRI connection as a recognized node of the network. The NT1 function provides the basic termination for the ISDN connection to your premises. An NT1 device also provides options for supplying power to the ISDN line and nonpowered ISDN devices.

The important thing to remember about the NT1 function is that you can have only one NT1 device for the termination of an ISDN connection. In most cases, you want your NT1 function handled by a stand-alone NT1 or NT1 Plus device.

Using an ISDN device — such as a remote access adapter card in your PC with built-in NT1 that doesn't include ports to connect other ISDN or analog devices — restricts use of an ISDN connection. Part III explains working with ISDN remote access devices.

Where to Start: NT1 or NT1 Plus?

NT1 or NT1 Plus devices perform the same network termination functions. Where NT1 and NT1 Plus devices differ is the types of equipment they allow you to connect to an ISDN line. The following sections explain which device to use, depending on the applications and devices you plan to support.

See Chapter 6 for more information on NT1 fundamentals and working with specific NT1 devices. Chapter 7 explains the essentials of NT1 Plus devices and Chapter 8 illustrates working with a specific NT1 Plus device.

About the NT1 Device

A stand-alone NT1 device works exclusively with ISDN-ready devices or terminal adapters for non-ISDN devices. It doesn't include the terminal adapter (TA) functions for connecting non-ISDN devices. Using an NT1 device, you can plug in S/T-interface devices. Figure 5-1 shows a typical ISDN line configuration based around an NT1 device.

The NT1 device offers a good solution if you don't plan to add analog devices, such as a telephone, fax, or modem to your ISDN line. However, not using an analog telephone with your ISDN line doesn't preclude you from working with voice communications — you can use an ISDN telephone. NT1 devices currently cost as little as $250.

A stand-alone NT1 device is the easier device to set up for your ISDN line because you don't need to configure it to handle non-ISDN devices. For example, to connect an analog telephone via an NT1 Plus device requires entering commands into the NT1 Plus device using the telephone keypad. The NT1 device also doesn't include any capabilities to manage your BRI connection. The devices that plug into the NT1 handle those functions. For example, a video conferencing system running on your PC controls the use of the B channels for voice and data. The NT1 acts simply as the conduit for the data.

The NT1 device typically includes one U-interface port for connecting an RJ-45 or RJ-11 cable. (Remember, you can use an RJ-11 connector in an RJ-45 jack.) An NT1 typically includes two S/T-interface ports with RJ-45 jacks.

Most ISDN devices include RJ-45 cables for connecting to an NT1. You can connect any of the following ISDN devices to S/T-interface ports as long as they don't have a built-in NT1.

- An ISDN adapter board in your PC that is part of a video conferencing system.
- An ISDN adapter card for remote access.
- A stand-alone ISDN bridge for remote access.
- An ISDN telephone.

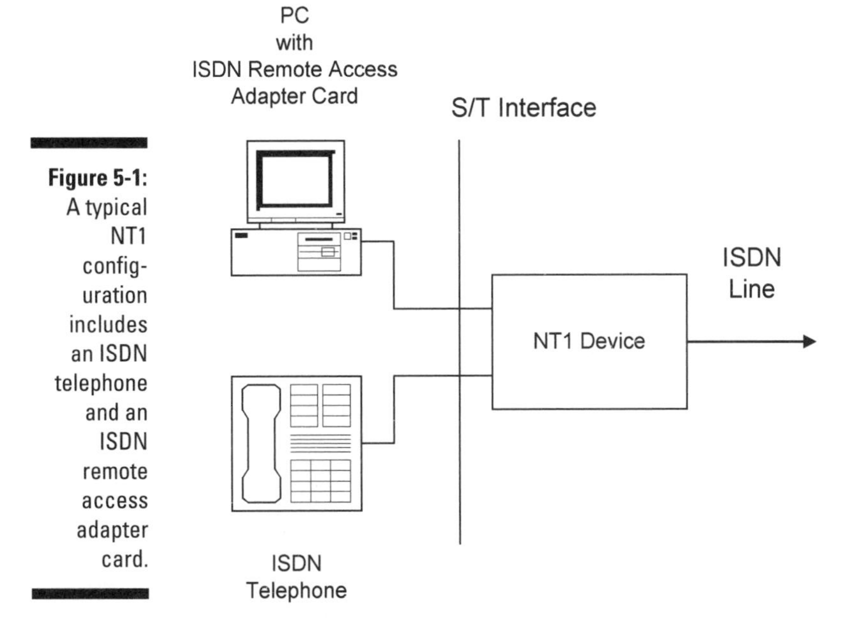

Figure 5-1: A typical NT1 configuration includes an ISDN telephone and an ISDN remote access adapter card.

About the NT1 Plus Device

An NT1 Plus device includes the same functions provided by an NT1, but also includes terminal adapter (TA) functions for connecting non-ISDN devices. If you want to use analog equipment on your ISDN connection or add a credit card scanner, you need to use an NT1 Plus device. Figure 5-2 shows a typical NT1 Plus configuration for an ISDN line. Determining which NT1 Plus device you want to use depends on the applications and devices you plan to use on your ISDN connection. NT1 Plus devices currently list between $350 and $1000, but prices are falling. The defining feature of an NT1 Plus device is that it includes an S/T-interface port plus an RJ-11 port for an analog device.

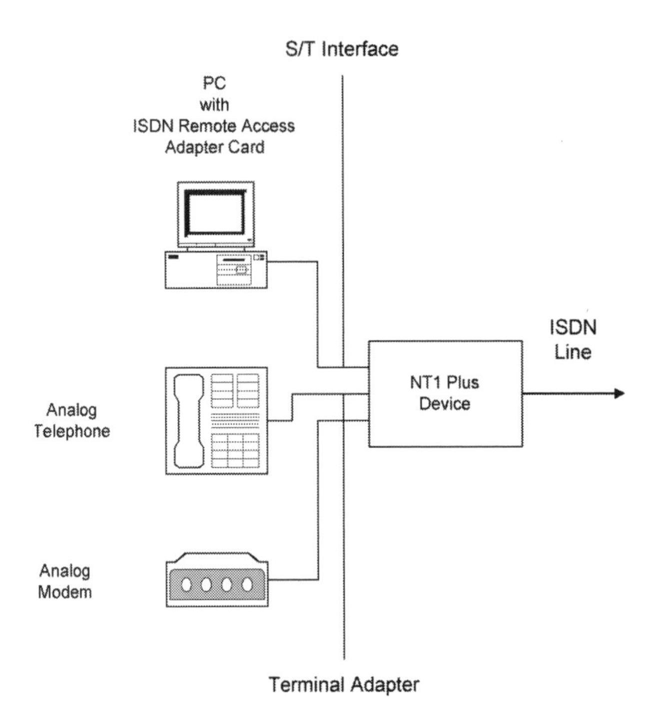

Figure 5-2: A typical NT1 Plus configuration includes an analog telephone, and modem, and an ISDN remote access adapter card.

NT1 Plus devices come in a variety of flavors for connecting a variety of non-ISDN-ready and ISDN-ready devices to an ISDN line. Most NT1 Plus devices include analog ports for connecting an RJ-11 connector from your telephone, fax, or modem in combination with other ports. Another common port on NT1 Plus devices is an RS-232 port for connecting your PC via a serial port for asynchronous data transmission. Some NT1 Plus devices also include ports for X.25 packet data transmission, which you can use to connect a credit card scanner.

The ISDN Voice Communications Revolution

ISDN is not only revolutionizing data transmission, it's also delivering a new generation of voice communication capabilities. ISDN delivers a dynamic array of call management features that, before ISDN, were only available in expensive business telephone systems. Chances are that at one time or another you've used these telephones in a large office setting. These telephones typically have a row of buttons at the bottom that illuminate when a call is active. To pick up a call, you press a flashing button. By pressing another button, you put a caller on hold. These telephones connect to telephone switching equipment housed at the company's premises. ISDN downsizes this system to the telephone set by eliminating the need for the expensive switching equipment. Instead, the functions previously handled by the on-site switching equipment are directly managed by the ISDN network and transmitted to an ISDN telephone or an analog telephone connected to an NT1 Plus device.

The Benefits of ISDN Voice Communications

ISDN offers powerful voice communication capabilities without new wiring or expensive switching devices. ISDN voice communication offers the following benefits and advantages.

- ISDN reduces the average time to make a connection from 12 to 30 seconds for an analog call to as little as three seconds. With SS7 implemented, which is explained in Chapter 2, this time is about a half second.
- ISDN lets you control the telephone network from your desktop via an analog telephone attached to an NT1 device or by using an ISDN telephone.
- ISDN lets you specify how you want to respond to calls based on whom the call is from or what the caller enters in response to recorded prompts.
- ISDN lets you bridge multiple calls into group conferences and change participants as needed using buttons instead of entering codes from the telephone keypad as is common in the analog world.
- ISDN delivers information about where the call is from and other data, which can be displayed on a telephone LCD.
- ISDN calls are more reliable than analog calls.

Working with ISDN Telephones

While you can use an analog telephone with ISDN via an NT1 Plus device, a telephone set designed for ISDN offers more efficient management of ISDN voice communications. The major advantage of ISDN telephones is the enhanced ability to control multiple calls using a separate button for each call. Figure 5-3 shows a typical ISDN telephone. An ISDN telephone also provides handy features for programming call management features into keys for quick access. A new generation of affordable ISDN telephones selling for under $300 is now coming online. An ISDN telephone can include the following items, depending on the model and the vendor.

- A U-interface or S/T-interface. The best solution is the S/T-interface version that you can plug into an NT1 device.
- A power supply (or it may require power from an NT1 device).
- Keys to program call management tasks, which can then be activated by the press of a button.
- An LCD display for conveying caller numbers and other call management information.
- Several buttons that handle multiple incoming calls from a single BRI line.
- An analog adapter to connect a fax machine or an RS-232 port to connect a PC.

Using an ISDN telephone, you can converse with one person on one call appearance. Another incoming call causes one of the other call appearances to flash a button on your telephone set. At the same time, the number of the calling party appears on the set's small LCD screen. You then have the choice of placing the first call on hold to answer the new one, bridging the new call onto the existing one for a conference call, or allowing it to be forwarded to a voice-mail system. If the caller abandons the call without leaving a message, an unanswered call log allows you to call the number back automatically.

Some of the less expensive ISDN telephones don't include an LCD screen, which prevents you from viewing an incoming caller's telephone number. Not having an LCD screen also makes programming the sets more difficult. Avoid these sets.

Figure 5-3: A typical ISDN telephone includes programmable buttons to handle a variety of call management features.

An ISDN Voice Communications Primer

If you plan to use your ISDN line for voice communication, whether using an analog telephone via NT1 Plus or an ISDN telephone, you need to understand some basic ISDN voice management fundamentals. The following sections explain the key elements of ISDN voice communications capabilities.

Centrex or Direct ISDN Voice Service

ISDN voice communication can be delivered via the ISDN network directly to your telephone, or you can use a Centrex-based system for multiple telephone environments. Centrex-based systems use the powerful call management features of the telephone switch itself. This type of service is set up and managed by the telephone company. It offers an affordable solution for small businesses. Delivering ISDN voice communications directly to your ISDN telephone is called distributed call management because your telephone controls your call management services.

ISDN voice services can be delivered to a PC, from which you can manage your telephone calls. Using an adapter card or NT1 Plus device with Microsoft Windows software, you can create a telephone system using a PC that acts as a telephone system switch device and performs other call management features. ■

ISDN's Supplementary Services

The collection of ISDN voice communication features are referred to as *supplementary services*. Supplementary services are the set of defined call management features you implement as part of your BRI service. You're already familiar with many of these features from the analog world. Call waiting, call forwarding, and conference calling are examples of supplementary services.

Supplementary services come in two flavors: local and end to end. Local supplementary service refers to services that require only the implementation at your end of an ISDN voice connection. End-to-end supplementary services require that the supplementary feature is implemented at both ends of the connection. The total set of supplementary services available at your premises depends on your telephone company and the services available from the central office serving your premises.

Directory Numbers and Call Appearances

The major advantage of ISDN for voice communication is its ability to handle multiple calls. Depending on the type of switch at the CO, you can have multiple directory numbers for each BRI line. Each directory number can have up to 64 different *call appearances* from a single BRI connection. Multiple call appearances of the same directory number let you handle more calls. These call appearances let you operate call waiting, with incoming calls coming into the same directory number. Unlike analog call waiting with its limit of two calls, an ISDN telephone can receive several incoming calls using buttons on the set.

Getting Keyed Up for ISDN

Setting up ISDN for voice communication involves telling the telephone company what supplementary services you want, although not all of these supplementary services are available at every CO. The telephone company configures their switch for these features. On your side of the connection, you configure your NT1 Plus device to work with the call management features you specified. This function involves working with key systems that define the functions of different buttons on a telephone for executing particular tasks. The key systems for ISDN voice communications are as follows. Electronic Key Telephone Service (EKTS), which is a National ISDN-1 standard. It supports call appearances based on the directory number for the voice terminal. You can have multiple directory numbers, but only one call appearance per directory number. However, using EKTS you can access a multitude of common ISDN call management features, such as Caller ID, automatic callback, conference calling, and call forwarding.

- CACH EKTS stands for Call Appearance Call Handling Electronic Key Telephone Service, a supplementary service offered for National ISDN-1. CACH EKTS lets you have multiple directory numbers and multiple appearances of each number if you use an AT&T 5ESS switch. There is a limitation of only one call appearance per directory number on the DMS-100 if the number is shared with another ISDN device.
- Feature Key Signaling lets you assign specific call management tasks to a given key on your ISDN telephone. Feature keys are identified by a unique number. For example, you might assign a key to call waiting so that when you press it, the switch at the CO knows you're specifying call waiting and responds accordingly. You can reassign keys as new services become available or if you no longer use various services.
- Keypad signaling is similar to the kind of actions you perform on an analog telephone for services such as call waiting, call forwarding, and conference calling. This type of call handling is typically incorporated into NT1 Plus devices that connect analog telephones to an ISDN line.

Getting Your Supplementary Services

When you establish your BRI service or at a later time, you can order your supplementary services. Different telephone companies sell ISDN supplementary services differently. For example, a telephone company might sell call appearances and supplementary services in units of ten, which you can then define as you want. You might order 10 units, take four units to create four call appearances, and use the remaining six units to add three-way conferencing, call waiting, call forwarding features, caller identification, repetitive dialing, and return last incoming call. You need to define your supplementary services so the telephone company can program its switch at the CO. Depending on your telephone company and the type of switch used at your local CO, you can have up to 64 call appearances and supplementary services. Other telephone companies offer a package of the most common supplementary services for a flat charge.

Common Supplementary Services

An extensive collection of supplementary services is available for ISDN. However, for most small businesses and individuals a core of services addresses most voice communication needs.

- Number identification supplementary services control the presentation of one party's ISDN number to the other party. Table 5-1 lists the most common collection of these services.
- Call offering supplementary services affect the connection and routing of calls. Table 5-2 lists the most common collection of these services.
- Call completion supplementary services affect the completion of incoming calls centering around call waiting. Table 5-3 lists the most common of these services.
- Multiparty supplementary services cluster around conference calling. Table 5-4 lists the most common of these services.

Table 5-1: Number identification supplementary services

Service	*Description*
Multiple Subscriber Number	Allows multiple telephone numbers to be assigned to a single BRI connection.
Calling Line Identification Presentation	Allows display of the caller's telephone number on an LCD display at the called party.
Calling Line Identification Restriction	Allows the calling party to prevent delivery of the calling ISDN number to the called party.
Connected Line Identification Restriction	Allows the connected party to restrict the display of its ISDN number to the calling party.

Table 5-2: Call offering supplementary services

Service	*Description*
Call Transfer	Allows transfer of an established call to a third party.
Call Forwarding Busy	Allows the network to automatically forward incoming calls to another number when your line is busy.
Call Forwarding No Reply	Allows the network to automatically forward incoming calls to another number if there is no answer on your line within a specified period of time.
Call Deflection or Call Forwarding Unconditional	Allows the network to automatically forward all incoming calls to another number even if your line is available.

Table 5-3: Call completion supplementary services

Service	*Description*
Call Waiting	Allows notification of an incoming call even when no information channel is available. You can then accept, reject, or ignore the incoming call.
Call Hold	Allows interruption of communications on an existing call and then reestablishing the connection.
Completion of Calls to Busy Subscribers	Allows camping on a busy telephone line until it's available. Once it's available, you're notified and the call is established.

Table 5-4: Multiparty supplementary services

Service	*Description*
Conference Calling (CONF)	Allows multiple users to simultaneously communicate with each other.
Three-Party Service (3PTY)	Allows placing an active call on hold and making a new call to a third party. You can then switch back and forth between the calls, join the calls together to form a three-way conversation, and split a three-way conversation back to separate calls.

Where to Now?

The next chapter provides more details about the workings of NT1 devices. It also shows you how to set up an NT1 device using products from several vendors. But before you decide to go the NT1 or the NT1 Plus route, read the next three chapters to get a feel for working with each type of device.

Chapter 6

Going the NT1 Route

In This Chapter

- Understanding Network Termination 1 (NT1) devices
- Working with ATI's UT620 NT1, Tone Commander's NT1U-220TC, ADTRAN's NT1 ACE, and Motorola's NT1D

As we've already learned, the NT1 is a pivotal piece of the ISDN connection. A stand-alone NT1 device lets you add S/T-interface devices to your ISDN connection. This chapter explains the workings of NT1 devices and takes you on a hands-on tour of working with three leading NT1 products.

The Workings of NT1 Devices

A typical NT1 device looks like a small modem. It usually has a series of LEDs on the front, with one U-interface port and two S/T-interface ports at the back of the unit. All the ports are RJ-45 jacks, though you can connect an RJ-11 cable into the U-interface port. You connect RJ-45 cables from your ISDN devices into the S/T-interface ports.

While the passive bus configuration for a BRI connection can support up to eight devices, in most cases you'll want to connect only two S/T-interface devices to an NT1. Adding more than two ISDN devices by using a Y connector gets complicated because you need to work with external resisters. NT1 devices include a switch to set the termination for your bus configuration, which is explain later.

NT1 devices ship with different power options depending on whether you need to power your customer premises equipment (CPE) from the NT1 or only the ISDN line. Most NT1 devices support both point-to-point and multipoint ISDN CPE configurations. They're also compatible with all CO switch types.

Powering Options

Your NT1 must supply power to your ISDN line. Beyond the power requirements of your ISDN line, which is minimal, you may need additional power supplied by your NT1. The CCITT defines the power for ISDN devices as Powering Source 2 (PS2). This provides dedicated power to S/T-interface devices that don't have their own power supply. If you're using ISDN devices that power themselves, such as ISDN adapter cards in a PC, you don't need the PS2 power supply. Some NT1 devices come only with the PS2 power option, which you can use even if your ISDN devices don't need power from the NT1. If you're using an ISDN device that doesn't include power, such as an ISDN telephone without its own power supply, you need the PS2 power supply with your NT1.

Things happen in an ISDN connection that can disrupt the operation of your NT1, and often the only way to clear it is to reboot the NT1. Unfortunately, most NT1 devices don't include a reset button. The only way to reboot an NT1 device without a reset button is to unplug it. Some NT1 devices include a reserve power option to allow time to shut down a connection in the event of a power loss.

Passive Bus Options

The term passive bus refers to the ability to connect multiple devices to a single BRI connection. The configuration of the passive bus combines the terminating resistors for all the devices connected to your NT1, which must add up to 100 ohms. Termination options for NT1s include 50 ohms, 100 ohms, and none (unterminated). How you set these options depends on the devices you're using with your ISDN line. You choose the termination option on the NT1 device that is added to any terminator resistor settings for the ISDN devices connected to the NT1. For example, if you're using a video conferencing system that includes a built-in 50-ohm terminator resistor, you set the NT1 termination option to 50 ohms, for a total of 100 ohms. The distances allowed for the different configurations vary depending on the NT1 Plus device.

A Sampling of NT1 Devices

A number of vendors offer NT1 devices. To give you a feel for working with NT1 devices, this chapter covers working with ATI's UT620 NT1, Tone Commander's NT1U-220TC, ADTRAN's NT1 Ace, and Motorola's NT1D.

Chapter 18 provides an extensive list of NT1 vendors and information about where you can buy NT1 devices. ■

Working with ATI's UT620 NT1 Device

The ATI UT620 is a compact NT1 device that lists for about $200. You can choose two different power supply options depending whether you need to power the ISDN devices or only the line. The power source is connected to a four-pin mini-DIN power connector located on the side of the UT620. If the UT620 is used with an SP62020 switching power supply, the UT620 can provide the CCITT standard Power Source 2 (PS2) power through the S/T-interface to the ISDN CPE. If the UT620 is used with the AD62010 power adapter, the UT620 does not provide the PS2 power to the ISDN CPE.

Understanding the Pretty Lights

The UT620 has three LEDs on its front panel that indicate the status of the power, S/T&U, and Loopback. The Power LED indicates that the power supply is functioning normally. When you first install the UT620, the S/T&U LED flashes for a few seconds, but then goes off. Here's how the UT620 communicates with you via its LEDs:

- If the U-interface isn't connected, the S/T&U LED flashes eight times per second.
- If the U-interface isn't activated, the S/T&U LED stays lit continuously.
- If the S/T-interface isn't activated, the S/T&U LED flashes once per second.
- If both the S/T- and U-interfaces are linked up, the S/T&U LED goes off.
- The Loopback LED lights during a CO loopback test with the UT620.

DIP Switch Settings

A block of four DIP switches is located on the bottom of the UT620. Use the DIP switches to configure the UT620 according to the settings in Table 6-1. In most cases, you'll choose the short passive bus option for DIP switch number 3.

Table 6-1: UT620 DIP switch configuration options

DIP Switch	*Function*	*On*	*Off*
1	S/T termination	Yes	No
2	S/T termination	Yes	No
3	Bus timing mode	Short passive bus	Point-to-point, or extended passive bus
4	PS2 selection	SP62020 switching power supply	AD62010 AC/DC power adapter

Installing the UT620

Installing the UT620 is an easy process that involves setting DIP switches, plugging in the power supply, and connecting the U-interface and S/T-interface cables. Here are the steps for installing ATI's UT620.

1. Set the DIP switches for your setup.
2. Connect the power supply to the UT620 and plug it into the AC power outlet. The Power LED lights.
3. Connect an RJ-11 or RJ-45 cable between the UT620 U-interface connector and the U-interface outlet at your premises.
4. Switch on the power to the ISDN terminal equipment. Then connect one end of an RJ-45 cable to an S/T port on the UT620 and the other end to the ISDN terminal equipment, such as an adapter card in your PC.
5. If you're connecting two ISDN terminals, repeat step four for the second device. The S/T&U LED may flash for a few seconds. The LED will go off when the communication paths have linked. When the LED goes off, the ISDN CPE is ready for communication.

Working with Tone Commander's NT1U-220TC

Tone Commander's NT1U-220TC lists for about $200. It includes the standard NT1 U- and S/T-interface ports. The NT1U-220TC also includes PS2 power for

ISDN devices not powered locally. The NT1U-220TC includes a handy feature that other NT1 devices don't — a power reset button. This button on the front of the unit lets you reboot the device. Figure 6-1 shows the NT1U-220TC.

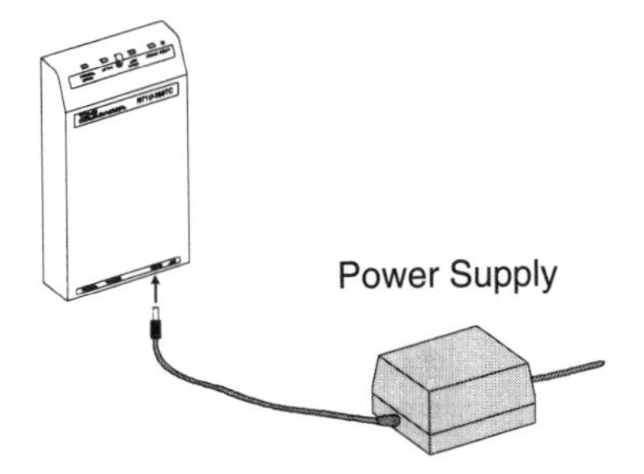

Figure 6-1: Tone Commander's NT1U-220TC includes a power supply for powering any ISDN devices connected it, and is the only NT1 device with a reset button.

Tone Commander also sells an NT1U-100TC unit, which is designed for ISDN applications that are powered locally. Unfortunately, it only includes one S/T-interface port. ■

Understanding the NT1U-220TC LEDs

The NT1U-220TC includes four LEDs on the front panel labeled TERMINAL ERROR, ACTIVE, LINE ERROR, and POWER. Here's how to interpret them.

- During power up of the NT1U-220TC, all four of the LEDs light for about a second.
- If the input power into the NT1U-220TC is too low, the POWER button flashes.
- If the TERMINAL ERROR LED lights, there is no active TE connected or there is a problem with the TE.
- If there is a problem with the U-interface, the LINE ERROR LED lights or flashes, and the TERMINAL ERROR LED flashes.

Setting Up the NT1U-220TC

Setting up the NT1U-220TC is similar to setting up other NT1 devices.

1. Set the TERMINATION switch for your configuration. Check the documentation of your ISDN devices to determine how you need to set the NT1U-220TC.
2. Connect the power supply to the NT1U-220TC and plug it into an AC power outlet. All the LEDs on the front panel light for about a second.
3. Connect one end of an RJ-11 or RJ-45 cable to the NT1U-220TC and the other connector to the U-interface outlet at your premises.
4. Switch on the power to the ISDN terminal equipment. Then connect one end of an RJ-45 cable to an S/T port on the NT1U-220TC and the other end to the ISDN terminal equipment, such as an adapter card in your PC.
5. If you're connecting two ISDN terminals, repeat for the second device.

Working with ADTRAN's NT1 Ace

ADTRAN's NT1 Ace includes the standard one U-interface and two S/T-interface ports. However, it comes in two pieces, the NT1 unit and a separate PS2 power unit, as shown in Figure 6-2. The ADTRAN NT1 Ace includes the PS2 power and 9.5 seconds of reserve power for bridging AC power interruptions.

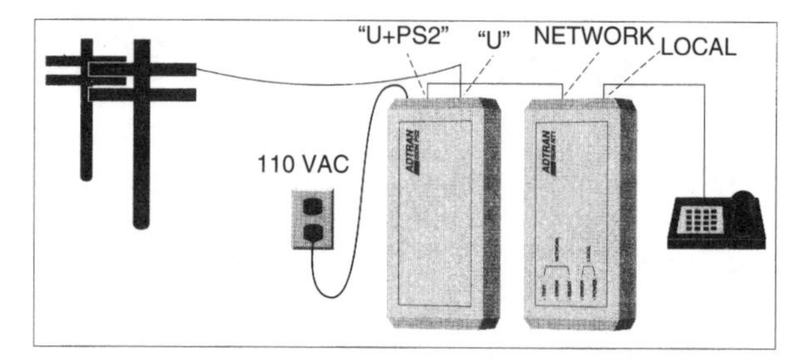

Figure 6-2: The NT1 Ace comes in two pieces and includes 9.5 seconds of reserve power in case of a power outage.

ADTRAN's NT1 Ace lists for $395, but in the near future they plan to release a similar version at a lower price. The new NT1 device will be a single unit instead of the two-piece NT1 Ace. ■

Making Sense of the NT1 Ace LEDs

Five LED status indicators on the NT1 Ace display the status of various parts of your ISDN connection, as described in Table 6-2.

Table 6-2: The NT1 Ace status indicators

When this LED lights	*It means*
Network Test	The network is performing a test
Network Error	The U-interface is not ready
Network Ready	The network is ready to place a call
Local Error	The S/T-interface device is not ready
Power	The NT1 Ace has power

Configuring the Local Bus

The two option switches located on the side of the unit are used to configure the local bus of the NT1 ACE. The switch labeled TERMINATION is used to select the local bus termination. The available options are NONE, 50, and 100. The second switch, labeled CONFIGURATION, has two possible settings. When set in the LONG position, the local bus is configured for extended passive bus. When set in the SHORT position, the local bus is configured for short passive bus. In most cases, use the SHORT option.

Setting Up the NT1 Ace

Setting up the NT1 Ace involves an extra step to connect the separate power unit to the NT1 unit, but it's still an easy process. You make the termination settings, plug in the power supply and connect your U- and S/T-interface cables. Here is how to set up ADTRAN's NT1 Ace.

1. Set the CONFIGURATION and TERMINATION switches for your configuration. Check the documentation of your ISDN devices to determine how you need to set the NT1 Ace.
2. Connect an RJ-11 or RJ-45 cable to the NT1 Ace PS2 unit's jack labeled U and the U-interface outlet at your premises.
3. Plug one end of the short RJ-45 cable into the jack labeled U+PS2 on the PS2. Plug the other end of the cable into the jack labeled NETWORK U on the NT1 ACE.
4. Plug the AC cord connected to the PS2 into a wall outlet that supplies 100 volts AC.
5. On the NT1 ACE, verify that the LOCAL POWER indicator is illuminated. The LOCAL ERROR and NETWORK ERROR indicators should also be illuminated. After about 15 seconds, the NETWORK ERROR indicator will go out.

6. Connect one end of an RJ-45 cable to an S/T port on the NT1 Ace and the other end to the ISDN terminal equipment. If you're connecting two ISDN terminals, do the same for the second device. After the TE powers up, the LOCAL ERROR indicator should go out. There may be some delay between plugging in the TE and the LOCAL ERROR indicator going out depending on the specific TE in use. Shortly after the LOCAL ERROR indicator goes out, the NETWORK READY indicator should illuminate.

The Motorola NT1D

The Motorola NT1D sells for about $200 and includes the standard two S/T-interface ports. Unlike other NT1 devices, it only lets you choose between no termination and a 100-ohm termination; it doesn't include a 50-ohm termination option. However, it does come with a power supply to power any ISDN devices not locally powered, such as an ISDN telephone. Figure 6-3 shows the Motorola NT1D device, which looks like a modem. Table 6-3 explains the functions of the six LEDs on the front of the unit.

Table 6-3: The functions of the six LEDs on the NT1D

LED	*Description*
SC (Sealing Current)	When on, this LED indicates the ISDN switch has bounced back a termination test voltage from the NT1D.
ACT (Activity)	When on, this LED indicates that a link between the terminal equipment and the ISDN switch at the CO via the NT1D has been established. If a disruption occurs between the U-interface and the ISDN switch, this LED flickers. If a disruption occurs between the S/T-interface and the terminal equipment, this LED blinks once per second. If a disruption occurs on both U- and S/T-interfaces, this LED goes off.
LB (Loop Back)	When on, this LED indicates the ISDN switch has sent a 2B+D loopback command to the NT1D.
LP (Local Power)	When on, this LED indicates the local AC power is active.
RP (Remote Power)	When on, this LED indicates the power at the remote site is functional.
RPR (Remote Power Reversed)	When on, this LED indicates the power at the remote site is not functioning properly.

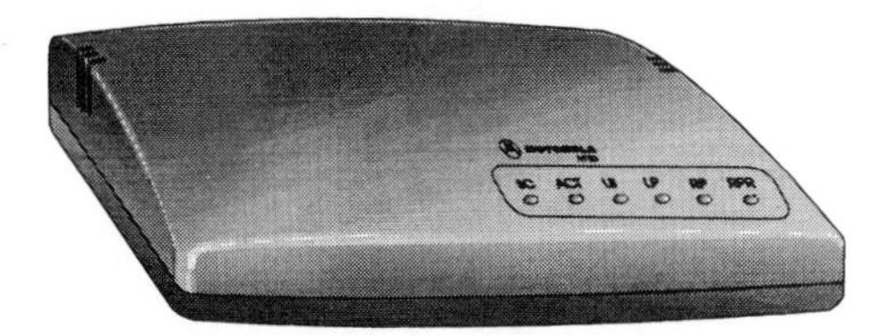

Figure 6-3: The Motorola NT1D looks like a modem.

End of the NT1 Trail

The NT1 solution is the easiest option to get up and running as long as you don't want any analog devices on your ISDN line. The next chapter explains the NT1 Plus option for adding analog devices such a telephone, fax, and modem to your ISDN line. If you decide to go the NT1 route, you can skip directly to Part III, which covers ISDN remote access solutions.

Chapter 7

Going the NT1 Plus Route

In This Chapter

- An NT1 Plus devices primer
- A survey of leading NT1 Plus devices
- A working tour of the ADAK 221 NT1 Plus

NT1 Plus devices include a network termination function like the NT1, but they also include terminal adapter capabilities to connect analog devices to your ISDN line. This chapter explains NT1 Plus fundamentals and provides guidelines for choosing the right NT1 Plus device.

NT1 Plus Device Primer

An NT1 Plus device includes the Network terminator 1 (NT1) function as well as terminal adapter (TA) functions. Currently, only a few NT1 Plus devices are on the market. An NT1 Plus device is defined as having an S/T-interface port and an RJ-11 port.

The typical NT1 Plus device requires a multipoint configuration with two directory numbers. You must configure one B channel for voice using the alternate voice/data configuration. The second B channel can be data only or alternate voice/data. You don't want to specify a voice-only B channel because you won't be able to use the channel for high-speed circuit data tasks, such as remote access or video conferencing. You also need to establish the supplementary services you plan to use for voice communications if you're connecting an analog telephone, as explained in Chapter 5. Different NT1 Plus devices support different call management services.

Using ISDN for Analog Communications

An NT1 Plus device may be your only option if you can only get two POTS lines to your premises. As you recall, a single telephone wire coming into your premises supports two separate telephone lines — two wire pairs.

If you need to convert one of your analog lines to ISDN, use the telephone line that is not your primary incoming voice number. Otherwise, when you're using an ISDN device that uses both B channels, it suffers a slowdown when it drops a B channel being used for data to answer the voice call.

Ports of Call and Other NT1 Plus Features

NT1 Plus devices come in a variety of port configurations for plugging in just about any type of device to your ISDN connection. The core ports that define an NT1 Plus device are at least one RJ-45 S/T-interface port and at least one RJ-11 POTS port for analog devices. Of course, every NT1 Plus device includes the U-interface port. An NT1 Plus may include the following ports.

- RJ-11 port for connecting standard analog devices, such as a telephone, fax, and modem. You can add multiple analog devices to an RJ-11 port.
- S/T-interface port, which uses an RJ-45 jack. This type of port allows you to connect ISDN devices, such as an ISDN adapter card for remote access or video conferencing.
- RS-232 port for asynchronous data transmission. This is the standard PC serial port that lets you connect your PC to communicate over ISDN without a modem.
- RS-232 port for synchronous data transmission for connecting a PC to an ISDN line to transfer data at the full 128 Kbps rate for the two B channels, or to support X.25 data transmissions for credit card scanners.

Any NT1 Plus device you get should support National ISDN National-1, AT&T 5ESS Custom, and DMS-100 switches. It should also provide support for automatic switching between 56 Kbps and 64 Kbps service. As you recall, long distance ISDN service typically drops to 56 Kbps because the signaling and bearer services are combined. If you use an NT1 Plus device that includes RS-232 ports for asynchronous and synchronous communications, make sure it includes the following features.

- Support for ISDN-based Internet access (or any TCP/IP network) via PPP/MP (PPP MultiLink Protocol). This new multipoint PPP standard for ISDN devices allows remote access devices to connect to the Internet using both B channels. It also allows for interoperability between different vendors' products.
- Support for Internet access via asynchronous communications using TCP/IP PPP.
- Support for the AT command set for communicating via a PC connected to an NT1 Plus device through a serial cable. Modems generally use the AT command set.
- Support for automatic switching between 56 Kbps and 64 Kbps service.
- Support for multiplexing, which allows for asynchronous data transmission over two B channels.
- Optionally, if you want to connect a point-of-sale device to an RS-232 port, make sure the NT1 device supports X.25 packet data transmission using the D channel, or connecting to host computers via a B channel over an X.25 network.

If you use an NT1 Plus device with an RS-232 port for circuit switched data communications, be sure it supports PPP MultiLink Protocol (PPP/MP).

NT1 Plus Devices vs. Digital Modems

NT1 Plus devices are defined by their inclusion of a combination of the S/T-interface port with a POTS port for plugging in a telephone, fax, or modem. Some NT1 Plus devices include an RS-232 port, but they all include the S/T-interface and the RJ-11 ports. Having an S/T-interface port lets you plug in ISDN equipment, such as an ISDN remote access device that doesn't include a built-in NT1 device. In most cases, you want to buy only S/T-interface versions of ISDN devices, such as ISDN telephones, remote access devices, and video conferencing systems.

Digital modems include RS-232 ports for asynchronous and synchronous communications, as well as RJ-11 ports for connecting a telephone or fax. However, what they don't include is an S/T-interface port for connecting other ISDN devices. These devices fall with the realm of ISDN remote access devices. As such, they're covered in Part III.

The Leading NT1 Plus Devices

Surprisingly, only a few NT1 Plus devices currently are on the market. The leading NT1 products are IBM 7845 NT Extended, ADAK Communications' ADAK 221, and ATI's Super NT1. Each product offers different options, but they all provide both analog and S/T-interface ports. The following sections describe these NT1 Plus devices.

Chapter 18 provides NT1 Plus vendor information. ■

The IBM 7845

The IBM ISDN Network Terminator Extended is a stand-alone unit that includes one U-interface port, one S/T-interface port, and one analog port. It includes power for both the ISDN line and PS2 power for nonpowered ISDN devices. The NT Extended also provides a standby battery for backup power to maintain analog telephone service during power outages. The NT Extended lists for $350.

The NT Extended uses one B channel for analog communication and one B channel for ISDN circuit switch data communication. If you're not using an analog telephone, you can use both B channels for circuit switch data. You plug your analog telephone into an RJ-11 port so you can take advantage of ISDN's supplementary services for voice communication. The NT Extended acts as a terminal adapter to make your telephone act like an ISDN telephone. For example, you can use such ISDN call management features as Caller ID, call waiting, and multiple call appearances. You can also connect a fax or modem to the NT Extended.

Chapter 8 takes you on a walk-through of IBM's NT Extended to give you a feel for working with an analog telephone connected to an NT1 Plus device. ■

The ADAK 221

The ADAK 221 includes two RS-232 synchronous/asynchronous ports, two RJ-11 analog ports, and an S/T-interface port. You can connect multiple POTS devices to the phone ports, a PC to one of the RS-232 ports, a credit card scanner to the other RS-232 port, and an ISDN-ready device to the S/T-interface port. The ADAK 221 incorporates high-speed B channel bonding capabilities to support data-intensive ISDN applications. It also includes built-in PPP and SLIP protocols for Internet surfing. In addition to robust data transport features, the ADAK 221 includes an extensive collection of call management functions for using ISDN's voice communications features with one or two analog telephones. It also supports X.25 communication for POTS devices and other public data network tasks. You can configure the ADAK 221 using a Touch Tone telephone keypad, or your can use the Microsoft Windows utility program that configures the ADAK 221 via a serial connection from your PC. Working with the ADAK 221 is explained later in this chapter.

ATI's Super NT1 Plus

Alpha Telecom's forthcoming Super NT1 device includes two analog ports and two S/T-interface ports. It also includes a small LCD screen for viewing ISDN voice communication features and for programming the device. This is the only NT1 Plus device with an LCD screen. The Super NT1 Plus also includes an internal power supply and a battery backup option for power loss protection for your analog communication. Attaching an analog telephone to the Super NT1 allows you to work with ISDN's supplementary services for call management.

Working with the ADAK 221

While at $799 the ADAK 221 sells for about twice the price of the IBM NT Extended, it provides more capabilities. The ADAK 221 includes a single S/T-interface port just like the IBM NT Extended, but it includes two RJ-11 ports for analog devices and two RS-232 synchronous/asynchronous ports. You can connect a single telephone or two telephones because the ADAK 221 supports using both B channels for voice. You can connect your PC directly to the ADAK 221 via a serial cable to use it as a modem for asynchronous communications to connect to online services that don't support ISDN. You can even connect a point-of-purchase device, such as a credit card scanner. If you only want to use one POTS for an analog telephone, you can use the other port for a fax machine or modem. Figure 7-1 shows the ADAK 221 ports and the devices that you can connect to them. You can purchase a battery to provide backup power for your analog communications in the event of a power outage.

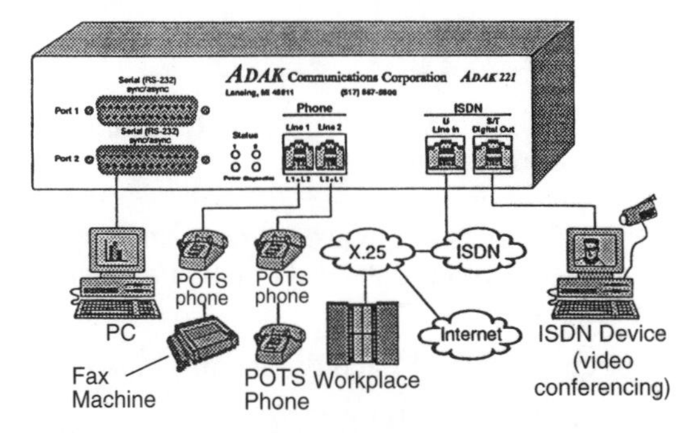

Figure 7-1: The ADAK 221 is a sophisticated NT1 Plus device that supports connecting a variety of devices to your ISDN line.

Because of the ADAK 221's expanded capabilities, it demands more from you to get it up and running. However, its Windows Configuration Utility makes configuring easier than using the telephone keypad.

Configuring Your ISDN Line for the ADAK 221

The first order of business is provisioning your ISDN line for working with the ADAK 221. For basic operation of the ADAK 221, the following sections cover the provisioning requirements for the leading telephone company switches.

AT&T 5ESS Custom

If your telephone company uses the AT&T 5ESS switch running Custom software, request the following configuration.

- Multipoint with two B channels configured for alternate voice/data, or one B channel for alternate voice/data and one B channel for data only if you want only one channel for voice.
- Two directory numbers — one for each B channel configured for voice.
- Supplementary Service Type=Type C.
- Calling Features: Conference, Transfer, Drop, and Hold.
- Four call appearances.
- Caller ID enabled, if Caller ID service is available in your area.

National ISDN 1 AT&T 5ESS

If your telephone company uses the AT&T 5ESS switch running National ISDN 1 software, request the following configuration.

- Multipoint with two B channels configured for alternate voice/data, or one B channel for alternate voice/data and one B channel for data only if you want only one channel for voice.
- Two directory numbers — one for each B channel configured for voice.
- Supplementary Service Type=CACH EKTS.

- Button Assignments: Conference=18, Drop=19, and Transfer=20.
- Four call appearances.
- Caller ID enabled, if Caller ID service is available in your area.

National ISDN 1 NT DMS-100

If your telephone company uses the NT DMS-100 switch running National ISDN 1 software, request the following configuration.

- Multipoint with two B channels configured for alternate voice/data, or one B channel for alternate voice/data and one B channel for data only if you want only one channel for voice.
- Two directory numbers — one for each B channel configured for voice.
- Supplementary Service Type=CACH EKTS.
- Button Assignments: Conference=18, Drop=19, and Transfer=20.
- Four call appearances.
- Caller ID enabled, if Caller ID service is available in your area.

Configuring the ADAK 221

Before you can use the ADAK 221 to make or receive calls, you must configure it. You can configure it using an analog telephone connected to the unit. You use the keypad on the phone to enter information to program the ADAK 221 firmware. But an easier way to configure the ADAK 221 is with the Windows Configuration Utility via a serial connection to the ADAK 221.

Installing the ADAK 221 Configuration Utility

The following steps explain how to install the ADAK 221 Configuration Utility.

1. Insert the ADAK Configuration Utility disk into your disk drive.
2. Choose File | Run from the Program Manager window.
3. Enter A:\SETUP and choose OK. The ADAK configuration program is installed in an ADAK Utilities group that is created on your system.

4. Connect your PC to the ADAK 221 via an RS-232 cable.
5. Double-click the Window Terminal icon in the Accessories group.
6. Choose Settings | Communications. The Communications dialog box appears.
7. Select the correct COM port in the Connector list, click 9600 in the Baud Rate setting, then choose OK. An asterisk prompt appears in the Terminal window. Leave the Terminal window open and proceed to the next section to configure the ADAK 221 for your ISDN line.

Configuring the ADAK 221 for Your ISDN Line

Once you've installed the configuration utility and connected your PC to the ADAK 221 unit via an RS-232 cable, you're ready to configure the ADAK 221. But before you do so, be sure you have all the ISDN line provisioning information from your telephone company, including your directory numbers and any assigned SPIDs.

The following steps explain configuring the ADAK 221 so you can use it for voice and attach an ISDN device in the S/T-interface port.

1. In the Windows Terminal window, enter PAD:PC_CONFIG and press Enter.
2. Exit the Terminal program and choose Yes to save the settings to a file.
3. Double-click the ADAK Configuration icon in the ADAK Configuration group. The ADAK configuration utility appears, as shown in Figure 7-2.
4. Click Switch Setup in the Main Select list. The configuration options for Switch type appear, as shown in Figure 7-3.
5. Choose the Service (switch software type) and the switch type.
6. Enter your area code in the Area Code field.
7. Click on Line 1 Setup. The Line 1 Setup configuration settings appear, as shown in Figure 7-4.
8. Choose the Supplementary Services setting for your ISDN line.
9. Enter the seven-digit number for your primary directory number in the Voice Primary Directory Number.
10. Enter the SPID number assigned by your telephone company, if any, in the Voice SPID field.
11. Click Line 2 Setup and repeat step 5 through step 10 for your second B channel.

12. Choose Option | Write Configuration to write your information to the ADAK 221. The Local Login Password dialog box appears.
13. Choose OK. The ADAK 221 Access dialog box appears. Choose OK. The Access Status dialog box appears.
14. Click Execute to write the configuration settings.
15. Choose File | Exit to close the ADAK configuration utility.

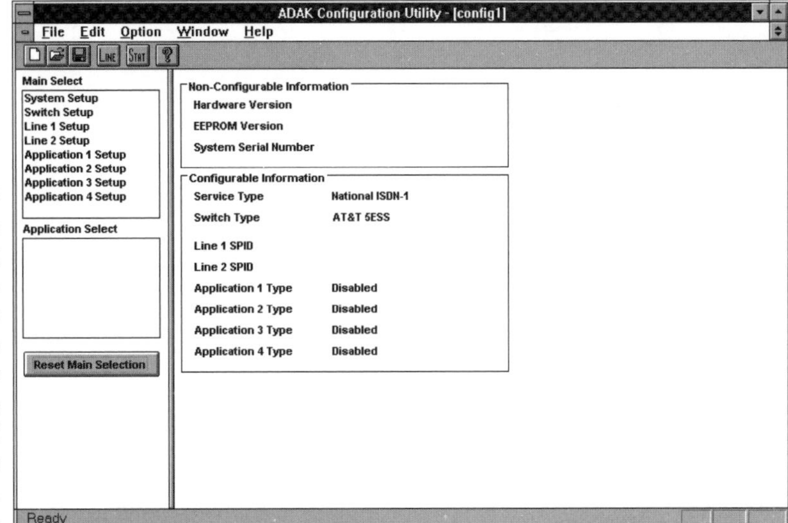

Figure 7-2: The ADAK configuration utility lets you easily configure the ADAK 221 from Windows.

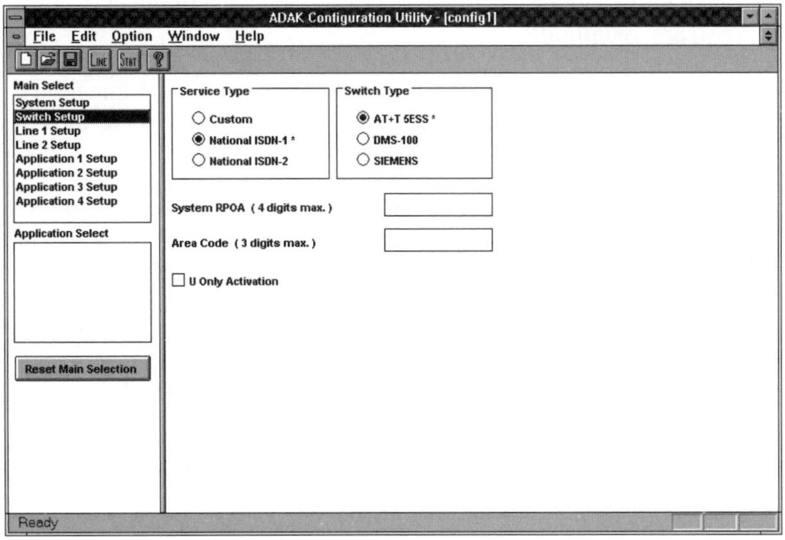

Figure 7-3: The Switch Setup screen lets you specify the telephone company switch platform and software.

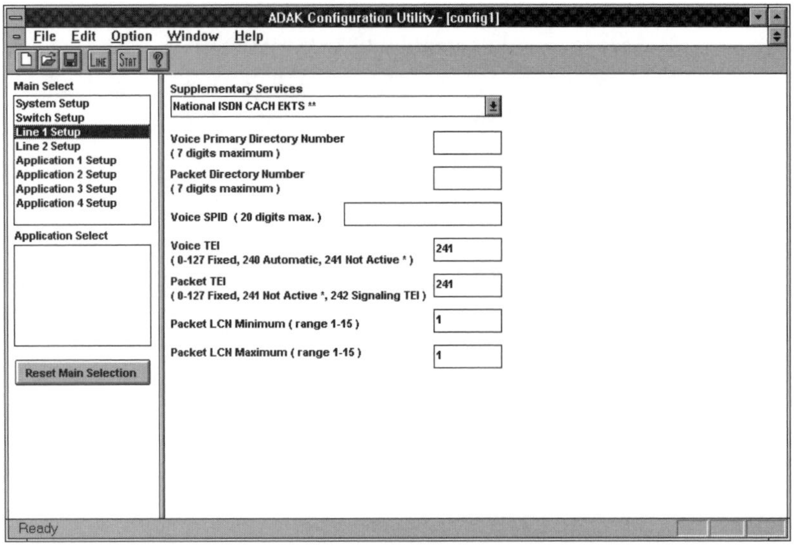

Figure 7-4: The Line 1 Setup screen lets you configure the primary B channel.

Using the ADAK 221 for Voice Communication

The ADAK 221 provides a wide variety of calling features using an analog telephone connected to a POTS port on the unit. If you configured your ISDN line to handle voice communication on both B channels, you can connect two analog telephones, each with its own telephone number. You can also use a two-line telephone.

The real power of using the ADAK 221 for voice communication via ISDN is that it transforms your telephone into a sophisticated multiline business telephone. The ADAK 221 can handle multiple incoming calls by letting you put calls on hold to switch between callers. For instance, if you're talking to one caller and need to place a second call, you put the first call on hold before dialing the second number. You can only talk to one caller at a time per telephone line, with all others remaining on hold until you retrieve them. The ADAK 221 can accommodate up to six calls per line. You must specify the number of call preferences for each number from the telephone company.

All ADAK 221 calling features are done with your telephone keypad. You must make all these key entries from the dial tone. This means that if you have a connected call, you must flash to put the call on hold before performing your entry. Two short beeps indicates that your entry is valid. If your entry is incorrect, you hear three beeps and a message stating why your entry was denied. Many calling features include voice messages generated from the ADAK unit to help you track what you're doing.

Placing and Receiving Single Calls

You place and answer single calls using the ADAK 221 the same way you do for analog service. To place a call, pick up the telephone receiver, listen for the dial tone, and dial the number. Answering a call is just as easy. Answer a call as you normally do by picking up the handset after the telephone rings.

Learning to Flash

The key mechanism for navigating among multiple calls or getting to a dial tone, is the flash. The flash is an action you perform to change between a dial tone and a call. To flash, press the receiver button on your telephone and release after approximately one-half second. If you release the receiver button too quickly or hold it down too long, the flash isn't correctly recognized. If you telephone has a special flash button, use it instead of the receiver button. You can place additional outgoing calls while other callers are on hold by doing a flash to get a dial tone, then enter an outgoing telephone number.

Working with Two Telephone Lines

You can use your ISDN line for two voice lines. Because each line has its own telephone number, you can connect a different telephone to each line or use a two-line telephone set so you can access both lines from a single set. Using a two-line telephone is easier because you just press the line 2 button to use the second line. To talk to the other line when using a two-line telephone, you need to switch lines using the appropriate controls for your telephone. Don't confuse the switching between telephone lines with using flash to navigate between calls on hold on a single line.

Handling Multiple Calls

Working with multiple calls involves using flash to put callers on hold and switching to another caller, or to get a dial tone. Multiple incoming calls on the same telephone number are referred to as call waiting. If the handset is off hook and a call arrives, you hear a single-beep call waiting. A flash puts the existing call on hold and answers the incoming call.

- To put a call on hold, use flash. To take a call off hold and resume the conversation, press flash to talk to the last call put on hold. If you hang up with calls on hold, the telephone rings. When you lift the handset, you're connected to the last call put on hold. When you flash a call to put it on hold, you hear a voice message say, "Call holding."
- To retrieving the last call put on hold, simply flash. If you have more than two calls holding and you want to talk to a specific call, retrieve the call by entering 2# followed by the call number in the order of your incoming calls. For example, to retrieve call number three, enter 2#3.
- To disconnect a call, hold down the receiver for at least one second. If you have a call on hold, the telephone will ring. When you pick up the receiver, you're connected to that call.

Redialing

Redial automatically dials the last outgoing number placed from the line. You can use the Redial command even if the original call did not successfully complete. To hear the last outgoing number, use the Redial Review command.

- To redial, from a dial tone, enter 3#.
- To review the last number dialed, enter 81#.

Calling Back the Last Caller

The Callback feature automatically recalls the last incoming call on the line. You can use this command even if you did not answer the call. To hear the last incoming number, use the Callback Review command.

- To use Callback, enter 4#. If the incoming telephone number is not available or the caller intentionally withheld the number, you hear a message saying, "Unknown number."
- To determine the last incoming call number, enter 72#

Adjusting the Volume

Adjust Volume sets the incoming sound level for all calls. The preset volume level is 4 on a scale of 0 (lowest) through 7 (highest). To adjust the volume level, flash to put the call on hold (if necessary). From the dial tone, enter 19# followed by the volume control number. For example to raise the volume to 7, enter 19#7.

Muting a Call

Mute silences your voice so the other party cannot hear you, but you can still hear him or her. This entry applies to the last call put on hold.

- To mute a call, flash to put the call on hold, then enter 18#. You hear a voice prompt saying, "Mute on."
- To turn mute off, flash to put the call on hold, then enter 18#. You hear the voice prompt saying, "Mute off."

Disabling Call Waiting

Disabling Call Waiting is used to turn off call waiting for the next outgoing call you make or the next call taken off hold. Call waiting resumes when the call is disconnected. This feature is helpful if you don't want to be interrupted during a call.

- To turn off call waiting (next call only) before placing a new call, enter 88#. When you hear the dial tone again, enter your telephone number or use speed dial.
- To turn off call waiting (next call only) while a call is already connected, flash to place the call on hold. From the dial tone, enter 88#, then flash to return to the call.

Forwarding Calls

Forwarding calls sends all incoming calls to any telephone number or Speed Dial Code you specify. When you establish call forwarding, your telephone doesn't ring; instead, all incoming calls ring at the new number.

- To forward all incoming calls to a specified telephone number, enter 10# followed by the telephone number to which you want to forward calls. For example, 10#5551212. You hear a voice message saying, "Forward to [telephone number]" before returning to the dial tone. As a reminder, you also hear this message before the dial tone if you pick up the telephone handset when forwarding is enabled.
- To turn off the call forwarding feature, enter 10##. You hear a message say, "Forward off."

Transferring Calls

The transfer command sends a call to a different destination after you answer it. There are two options for transferring a call. You can transfer the current call to a new telephone number or you can transfer a call that's on hold.

- To transfer the current call to a new telephone number, flash to put the call on hold, then enter 6#*telephone number*#. Hang up or listen for the dial tone.
- To transfer the last on-hold call, flash to put the new call on hold, then enter 6##. You must have at least two connected calls to use this transfer method.

Working with Conference Calls

The conference call feature joins you and at least two other parties into one call so all parties can converse with each other. You have two options when creating a conference or adding parties to a conference. You can dial a new telephone number to join you and the last call put on hold, or you can take the last two calls put on hold and conference them with you. Only one conference is allowed per telephone line.

- To conference the call you are currently talking to and add a new telephone number, flash to put the call on hold, then enter 7#*telephone number*#. You automatically return to the conference.
- To conference with the last two calls put on hold, flash to put the call on hold, then enter 7##.
- To drop a party in a conference call, including if a party you're calling doesn't answer, enter 8#.

Working with Speed Dialing

Speed dialing allows you to quickly place outgoing calls by dialing a two-digit Speed Dial Code followed by a pound sign (#). Before using Speed Dial, you must set the telephone number that will be dialed when you later use the Speed Dial Code.

- To create a speed dial entry, enter 80#*Speed Dial Code Number*#, where the *Speed Dial Code Number* is a number between 40 and 59. You're prompted to enter the telephone number. Enter 1 followed by the telephone number as you would dial it, including any prefix and area code.

- To erase a speed dial entry, enter 80#*Speed Dial Code Number*#, where the *Speed Dial Code Number* is the entry you want to erase, then enter 2.
- To place a speed dial call, enter the Speed Dial Code number followed by #. For example, entering 45# dials the telephone number you specified for Speed Dial code 45.

Where to Next?

You now have a handle on NT1 Plus devices, including experience using the ADAK 221. The next chapter takes you through working with the other leading NT1 Plus device — the IBM NT Extended. This NT1 Plus device includes many, but not all, of the features of the ADAK 221.

Chapter 8

Using the IBM 7845 NT Extended

In This Chapter

- How to install and set up the IBM NT Extended
- Using the NT Extended voice communication features

The IBM 7845 NT Extended, or NT Extended for short, is currently the most affordable NT1 Plus device available. Listing at $350, the NT Extended allows you to plug in analog devices as well as an S/T-interface device using an RJ-45 port. This chapter takes you on a hands-on journey through setting up and working with the NT Extended to get a feel for how an NT Plus device works.

Getting Started with the IBM NT Extended

The NT Extended is a stand-alone unit shown in Figure 8-1 that includes one U- and one S/T-interface port, and one analog port, as shown in Figure 8-2. It includes power for both the ISDN line and PS2 power for nonpowered ISDN devices. The NT Extended also provides standby battery backup power to maintain analog telephone service during power outages.

The NT Extended uses one B channel for analog communications and one B channel for ISDN circuit switched data communication, or you can use both B channels for circuit switched data. Using the NT Extended for analog voice communication through a B channel is referred to as *extended analog or extended mode*. With the NT Extended's extended analog features, you can use your analog telephone through an ISDN line and take advantage of ISDN's supplementary services. For example, you can use ISDN call management features such as call waiting, multiple call appearances, and more. You can also connect a fax or modem to the NT Extended.

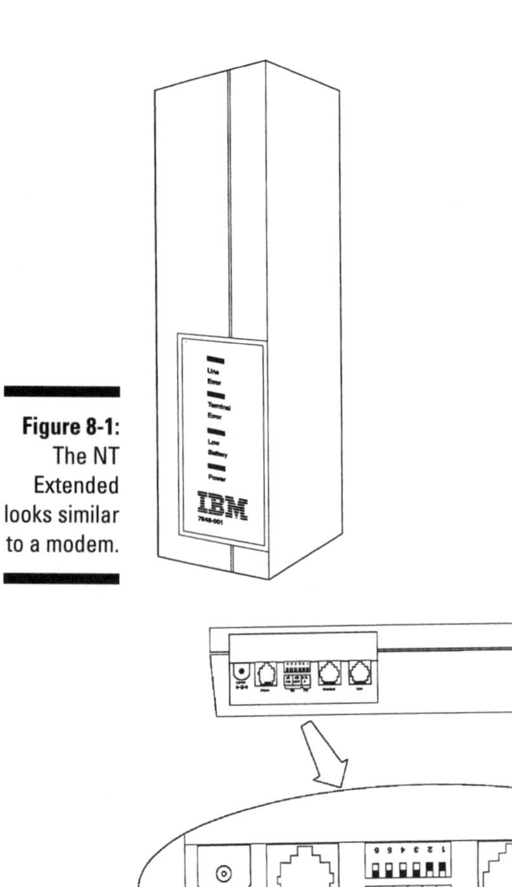

Figure 8-1: The NT Extended looks similar to a modem.

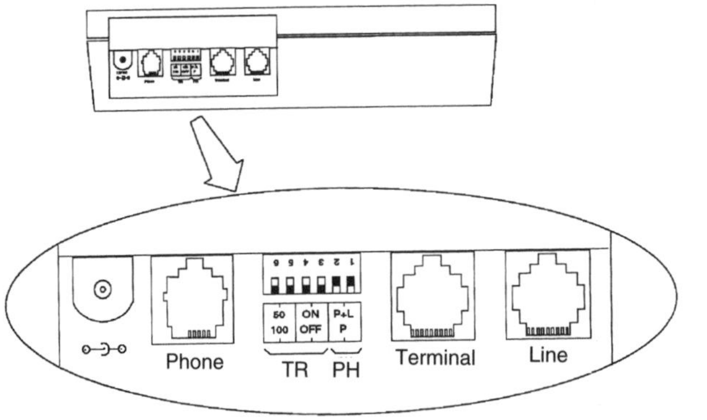

Figure 8-2: The NT Extended's back panel includes one U-interface port, one S/T-interface port, and one analog RJ-11 port.

Ordering ISDN Service

You must use at least two directory numbers if you want to use both an ISDN device and an analog device. Calls placed to the first number are answered by the ISDN device while calls placed to the second number cause the analog telephone to ring. The 7845 lets you use up to five call appearances for each directory number.

You need to provision your ISDN line for the NT Extended by ordering the following configuration from your telephone company. The telephone company incorporates the configuration options into their switch.

- ISDN BRI service (2B+D) multipoint
- Terminal Type C
- EKTS with CACH
- If the telephone service provider requires a key set name, use NTEXTENDED.
- Two directory numbers, one for circuit-switched data and one for voice
- If the telephone service provider uses a DMS-100 switch, either configure each B channel for voice/data, or configure one B channel for alternate voice/data and the other only for data. For 5ESS, configure each directory number on the multipoint line as each device requires. The 7845 uses one channel for voice, while the rest of the configuration depends upon the requirements for each device.
- At least one additional call appearance, but you may want to add several more to handling multiple incoming calls. The NT Extended supports up to five call appearances.

If you're connecting an analog telephone to the NT Extended, you must define your calling features before provisioning your ISDN line. See the section entitled Working with NT Extended's Calling Features later in this chapter to determine the key options you want. If you only want to use a fax machine or modem for your communication, you don't need to deal with setting up call provisioning features for your SPID.

One other configuration setting that is necessary for the telephone company is to assign unique keys to several call management features. Table 8-1 specifies the required key assignments for the NT Extended.

Table 8-1: The NT Extended key assignment functions

Key	*Assignment*
1	Primary directory number
2	Additional call appearance of primary directory number
3-7	Additional call appearances of primary directory number beyond the first one
11	Secondary directory number
12	Tertiary directory number
21	Three-way or six-way conference
22	Drop
23	Hold (if included in conference features package)

Setting Up and Working with the NT Extended

The NT Extended's Line port at the back of the unit is for plugging in your ISDN BRI line. You can plug an RJ-11 or RJ-45 cable from the U-interface into the NT Extended. The phone port lets you plug in any analog device using an RJ-11 jack. The terminal port allows you to connect an S/T-interface device, such as a video conference system.

You configure the NT Extended with the keypad on your Touch Tone telephone, which you connect to the NT Extended's phone port. You use the keypad's 0-9, *, and # keys to enter commands to configure the NT Extended. Before you can start using the NT Extended, you must tell it your primary directory number, switch type, and any assigned SPIDs. The NT Extended stores these configuration settings permanently even if the AC power and the battery backup power are lost.

Establishing Termination Resistance

You must establish the termination resistance of the NT Extended. The NT Extended supports no terminating resistance, 50-ohm termination, and 100-ohm termination. If you're using an S/T-interface device that incorporates a 50-ohm termination resistor, you must set the NT Extended to 50 ohms, which gives your connection a total of 100 ohms of resistance. Remember, each BRI line must terminate with a total of 100 ohms.

To set the termination resistance for the NT Extended, you change DIP switch setting located at the back of the unit. The NT Extended has six DIP switches, of which you use switches 4–6 to set the termination resistance. Here are the settings.

- To specify no termination resistance, turn switches 4–6 to the off position (down).
- To specify 50-ohm termination resistance (the default), make sure switches 4-6 are switched on (up).
- To specify 100-ohm termination, turn switches 3 and 4 to the on position, and switches 5 and 6 to the off position.

Entering Configuration Commands

To enter commands for setting up the NT Extended to work with your ISDN line, the NT Extended must be in Standard mode, which is the default setting. Standard mode is one of the two modes of operation of the NT Extended. The

Standard mode allows only circuit data through the two B channels. The Extended mode allows you to use an analog telephone on one B channel and an ISDN device on the other B channel.

When you place the NT Extended in Extended mode, you hear a dial tone and are ready to use the telephone functions. If you hear a reset tone (two short beeps of different pitch every three seconds) or dial tone following the entry of a command, this means the command is accepted. An error tone — a continuous sequence of short, alternating-pitch tones — indicates a bad command entry.

To switch the NT Extended between Standard and Extended modes, do one of the following:

- To place NT Extended in Extended mode from Standard mode, enter *#9.
- To place NT Extended in Standard mode from Extended mode, enter *#0.

Setting Up the NT Extended for Your ISDN Line

Before you can start working with the NT Extended, you must enter commands to configure it for your ISDN line. The following steps explain how to do so.

1. Take the handset off your telephone set. When the handset of the analog telephone attached to the NT Extended is first taken off-hook, you should hear a reset tone.
2. Set the Primary Directory Number by entering *98, your ten-digit telephone number, and #. For example, enter *985085559021#. After successful entry of the setup command, you should hear the reset tone again.
3. Set the Network Switch Type by entering *97, the switch value, and #. The switch values are listed in Table 8-2. For example, enter *971# for an 5ESS NT1 switch.
4. Set the SPID numbers provided by your telephone company (if any) by entering *99, SPID number (1-25 digits), and #. For example, enter *9950855512120000#.
5. Enter *#9 to place the NT Extended in Extended mode. If successful, you should hear a dial tone ready for you to make a call.

Table 8-2: Switch values for configuring the NT Extended

Switch type	*Switch value*
DMS-100 NI1	1
5ESS NI1	4
5ESS Custom	5

Restoring Factory Settings

You can return the NT Extended to all its original factory settings by entering *9*9#. Following this command, the reset tone replaces the dial tone. In this setting, the NT Extended operates in Standard mode.

Understanding Distinct Number Ringing

Three ring cadences are possible, depending upon the number of directory numbers assigned to the analog telephone system. Calls placed by a remote caller to the primary directory number ring with cadence 1. If you subscribe to a second or possibly a third directory number, calls to these numbers ring with cadence 2 or 3, respectively.

Cadence l: Long

Cadence 2: Short, short

Cadence 3: Short, long

Mastering Hook Flashing

The NT Extended supports three types of hook flashes. A hook flash is the quick depression of the switch hook. The three types of hook flashes supported by the NT Extended are:

- *Short Hook Flash.* This is a temporary depression of the switch hook that lasts less than one second. You use the short hook flash to switch between calls. Some phones have a flash key that you can use to generate a short hook flash.
- *Long Hook Flash.* This is a temporary depression of the switch hook for longer than two seconds. Using a long hook flash is the same as if you placed the telephone on-hook for a prolonged period of time, then took it off-hook. Under some conditions, when you place the telephone on-hook for a prolonged period of time, the telephone will ring back.
- *Double Hook Flash.* This involves two short hook flashes, back to back within a two-second period. You use the double hook flash in a variety of cases in which you want to get a dial tone and a single hook flash would cause you to connect with another held or waiting call.

Getting Toned Down

The NT Extended generates a collection of tones that signal a particular event. Many of these tones are similar to those you hear on analog telephones. Table 8-3 lists and describes each of the tones generated by the NT Extended.

Table 8-3: Tones you hear from an analog telephone attached to the NT Extended

Tone	*Description*
Reset Tone	Two short beeps of different pitches repeating every three seconds. You hear this tone when the NT Extended is in Standard mode or when you go off-hook.
Dial Tone	A steady tone indicating that the NT Extended is operating in Extended mode and is ready to make and receive calls.
Ring in Progress	The normal tone that indicates the number dialed is ringing at the remote location.
Busy	The normal tone that indicates the number dialed is in use.
Off-Hook Timeout	A very fast and loud busy tone that sounds when you don't complete a dialing command or enter the first digit of a telephone number within 20 seconds of the time you went off-hook. If the phone continues to stay off-hook with no activity for 30 more seconds, this tone gets louder.
Error Tone	A continuous sequence of short, alternating-pitch tones. You can terminate the error tone by going on- hook or by using a hook flash.
Call Waiting	You hear the call waiting signal repeated every 10 seconds. Different tones indicate to which directory number the call has been directed. Directory number 1 is a long, high-pitch beep. Directory number 2 is a long, low-pitch beep followed by a short, high-pitch beep. Directory number 3 is a short, high-pitch beep followed by a long, low-pitch beep.
Low Battery Warning	A very short, high-pitched beep repeating every 10 seconds. This warning carries over any activity occurring while the telephone is off-hook and indicates that the battery backup charge is low. Only 10 minutes of off-hook capacity remains in the battery. This signal should occur only when local AC power is interrupted.

Working with the NT Extended's Calling Features

The NT Extended supports a large variety of custom calling features. To use these features you first must order call features from your telephone company. You then enter the commands for the NT Extended's call features when it's in Extended mode and you hear a dial tone. The following sections explain working with NT Extended's basic call features, which include:

- Speed dialing phone numbers using two-digit locator numbers.
- Redialing the last number you dialed.
- Attack dialing, which let's you camp on a busy line until the other party hangs up. Your telephone then automatically connects to the number.
- Returning the last incoming call that didn't connect because your line was busy.
- Establishing call blocking to block incoming calls from a small database of numbers.
- Working with call waiting to manage multiple incoming calls.
- Managing multiparty conference calls.

To distinguish between telephone numbers and commands, all NT Extended commands start with the * key. Some commands that are variable in length must be terminated with a #. During an active call, the NT Extended disregards any key you press except for the switch hook. If you get an error tone while entering a telephone number or a command, hang up or use a hook flash to terminate the error tone.

Speed Dialing and Redialing

You can establish up to 30 speed dialing numbers in the memory of the NT Extended. These numbers are identified as numbers 01 through 30. Speed dialing numbers can be of any length up to 21 digits.

- To enter a speed dialing number, enter *95, a two-digit number between 01 and 30, the telephone number, and #.
- To remove a speed dialing number from memory, enter *95, a two-digit number between 01 and 30, and #.
- To use speed dialing, enter * followed by a two-digit memory number between 01 and 30.

For example, to assign 1-508-555-5555 to speed dial number 05, enter *95 05 1508555555 #. To speed dial this number simply enter *05 on your telephone keypad. If you later want to remove this speed dial assignment, enter *95 05.

When you enter *00, the NT Extended redials the last number you dialed, whether manually or with speed dialing. You enter this command when you hear the dial tone.

Charging Ahead with Attack Dialing

Whenever you dial a number that is busy, the number is stored in the NT Extended's memory. You can tell the NT Extended to repetitively dial the stored number every 30 seconds until the called number is no longer busy, or you take the telephone off-hook to make another call. Any incoming calls that arrive while the repetitive dialing is in progress hear a busy tone.

To commence attack dialing after receiving a busy tone, do a hook flash to obtain a dial tone, enter *66, then hang up the telephone. When the NT Extended receives an indication that the called party's telephone is ringing, your analog telephone rings with a special alert cadence. What occurs next depends upon who picks up the telephone first. The possibilities are as follows.

- If the called party picks up the telephone first, they hear nothing until you pick up the telephone.
- If you pick up the phone first, you hear a ringing signal indicating that the called party's telephone is ringing.
- If the called party hangs up the telephone before you pick up the telephone, your telephone will stop ringing.
- If you do not answer the ringing telephone within 30 seconds, the NT Extended terminates the call and the ringing stops.

Returning the Last Incoming Call

You can have the NT Extended automatically return the last incoming call even if you were on another call. For example, if you have a caller that's been put on hold and they hang up, you can automatically call them back. This feature requires the availability of Calling Number Delivery from your telephone company. This feature is not available in all states because of privacy issues. If your service includes Caller ID, the NT Extended stores the caller's number in memory whether the call is answered or not.

To activate the feature that returns the last call, enter *57. The first time you execute this command, the calling number is dialed with only the last seven digits. This is assuming that the area code as specified in your primary directory number and the area code of the calling party is the same. If this is not the case, this step is skipped. If the call goes through successfully, no additional steps are required. However, if the call doesn't go through, do a long hook flash, then press *57 again. The number is dialed with the area code and a 1 in front of it. If the call still doesn't connect, do a long hook flash and press *57 again. The third time, the NT Extended dials the number with the area code but without the preceding 1.

Establishing Call Blocking

The call blocking feature allows you to store up to 10 telephone numbers in the memory of the NT Extended for blocking calls from those numbers. After establishing call blocking, anyone on that list will receive a busy signal if they try to call you. The call blocking command is dependent upon the availability of, and subscription to, the ISDN Calling Party Delivery feature from your local telephone company. The various call blocking commands are as follows:

- To establish a call blocking number, enter *94, a two-digit memory location between 01 and 10, a 10-digit telephone number, and #.
- To remove a previously entered call blocking number, enter *94, the applicable two-digit memory location between 01 and 10, and #.
- To activated call blocking for all the numbers you've specified, enter *60.
- To deactivated call blocking for all the numbers you specified, enter *80.

Working with Call Waiting

The NT Extended provides call waiting services that include features for holding and retrieving multiple incoming calls. You can hear the call waiting signal at any time during an active call.

You can place a call on hold to either answer a waiting call or to place an outgoing call. If you ordered additional call appearances for your ISDN line, you can manage more than two calls at a time. You use the hook flash to manage waiting calls as follows:

- A short hook flash (tapping the switch hook for less than one second) lets you hold active calls or connect nonactive calls. A short hook flash always places an active call on hold. If a call is waiting, you connect with that waiting call. If no call is waiting but there is a call on hold, you connect with the oldest call on hold. If there is no call waiting and no calls are on hold, you get a dial tone.

- A double hook flash (two short hook flashes back to back) connects you to an available call appearance for an outgoing call. The active call at the time you perform the double hook flash is placed on hold. Any waiting calls at the time of the double hook flash are put on hold. Calls that were already on hold at the time of the double hook flash remain on hold.
- A long hook flash (pressing the switch hook for more than a few seconds) disconnects the active call. If a call is waiting at the time of the long hook flash, you're connected to that waiting call. It no calls are waiting but calls are on hold, the NT Extended connects you to the oldest call on hold. If there are no calls waiting or on hold, the result of a long hook flash is a dial tone.
- Going on-hook. While there are waiting calls or calls on hold, you may go permanently on-hook. If you do, a ring-back occurs and, when you answer the telephone, the NT Extended connects you with the same party as if you had done a long hook flash.

If you don't want any interruptions during an outgoing call, you can suppress call waiting for the duration of an outgoing call. You can't suppress call waiting during incoming calls. To suppress call waiting on an outgoing call, pick up the phone, enter *70, then make your call. After you hang up, call waiting is automatically reactivated.

Working with Conference Calls

To produce a conference call you must subscribe to either the ISDN 3-Way Conference facility or the ISDN 6-Way Conference facility with your telephone company. During a conference call, all incoming calls not part of the conference receive a busy signal. To establish a conference call, do the following.

1. Start with an active call and place it on hold by a short hook flash. This gives you a dial tone.
2. Place a second call and wait for the answer. If there is no connection, go to the next step.
3. Perform a double hook flash to place the second call on hold along with the first call.
4. Press **1. At this point, each of you is a part of the conference.

You can add other parties (a third party if you have three-way conferencing or up to four additional callers if you have six-way conferencing) to a conference call by first placing the conference on hold by performing a short hook flash. Dial the new party, and after that party has answered, add the new party to the conference by doing a short hook flash. To drop the last party that you added to a conference, break out of the conference by performing a short hook flash, press **0, then perform another short hook flash.

A conference call ends if all of the remote parties hang up. It also ends if you perform a long hook flash or permanently go on hook while you are part of the conference.

Past the First Fork on the Road to Digital Delights

You now have a clear view of the two routes you can take to terminate your ISDN connection. You're also aware of your ISDN voice communications options. From this vantage point you're ready to start working with specific ISDN applications, starting with remote access. Continue on to the good stuff. . . .

Part III

Remote Access via ISDN

"IT'S ABOUT A GROUP OF YOUNG, INEXPERIENCED GERMAN PROGRAMMERS ALL CRAMPED TOGETHER INSIDE THIS HOT, TINY COMPUTER ROOM FIGHTING THE VIRUS OF THEIR LIVES DURING THE EARLY YEARS OF NETWORK COMMUNICATIONS."

In This Part...

Remote access is the number one ISDN application. So you now will learn how to harness ISDN for high-speed remote access. You'll learn how to connect your PC via ISDN to telecommute to your office LAN, connect to the Internet, and transfer files at rocket speed. The following chapters explain your remote access CPE options and sort through them to help you choose the right solution. They guide you through developing a plan that gives you the most flexibility in remote access capabilities. You'll also experience the process of installing, configuring, and using specific remote access solutions.

Chapter 9

Using ISDN for Remote Access

In This Chapter

- Understanding the benefits of ISDN for remote access
- Learning the options available for remote access via ISDN
- Developing your remote access game plan

Remote access is the workhorse of all ISDN applications. Whether you're connecting to the office for telecommuting, using an online service, surfing the Internet, or accessing any host computer, ISDN is the way to go. Because of the power ISDN delivers for remote access, it's no surprise that you can find a number of CPE options. This chapter presents a foundation to help you evaluate remote access options for ISDN connectivity.

ISDN-Based Remote Access

The expanded bandwidth of ISDN translates into moving larger amounts of data in shorter periods of time. You can use ISDN for all kinds of remote access activities.

- For telecommuting, ISDN offers an affordable, high-speed connection from your home PC to the company LAN using a seamless ISDN connection to work with applications and files. The speed at which ISDN delivers data makes you feel like you're working right at the office.
- For Internet cruising, ISDN offers a high-speed connection for working and playing on the Internet. ISDN moves data at speeds that make the popular World Wide Web come alive at your desktop. Not only does ISDN speed up your Web travel adventures, the reliability of your Internet connection improves dramatically.

- For connecting to online services, ISDN offers an attractive route for working faster online. You can perform all kinds of daily tasks on services such as CompuServe, America Online, and PRODIGY at unprecedented speed. Most online services plan to offer ISDN connections in the near future.
- For providing information online, ISDN access to multimedia and other data-intensive resources creates new opportunities for delivering information. For example, a real estate database can include pictures, layouts, even videos that people can browse online or quickly download.
- For moving large files, ISDN lets desktop publishers, multimedia developers, software companies, and a host of others deliver their software products economically and efficiently.

The Limits of On-Demand ISDN (Time Is Money)

ISDN delivers an affordable, on-demand remote access system for PCs. The key term here is on demand. ISDN is a usage-based service for which you pay according to the time you use it. ISDN is fine for making calls to perform specific tasks on an as-needed basis, but it's not a solution if you plan to connect to remote sites for longer than a few hours each day. You need to check out tariffs to calculate what your remote access connections will cost. However, remember that ISDN lets you work a lot faster than a modem. If you need more than a few hours of connection time each day, consider leasing a private digital line — but be aware it will cost several hundred dollars per month.

Remotely Accessing a LAN via ISDN

The easiest way to connect to an Ethernet-based LAN is with an Ethernet-based ISDN adapter card in your PC. Ethernet is the protocol used by most PC-based networks. With both systems running compatible networking software, you can connect to the remote network via ISDN. When you specify a host computer from your PC, the connection is automatically made by your ISDN adapter to that computer at the remote site. From your vantage point, it looks like you're sitting at a node on the remote network. A node is any PC or other device connected to a network and capable of communicating with other network devices. ISDN is the conduit that operates seamlessly between your PC running as a network node and the local area network operating at the other end — which can be down the block or on the other side of the planet.

Surfing the Internet via ISDN

Connecting to the Internet via ISDN operates in the same manner as connecting to a LAN. You use TCP/IP software on your system to connect via ISDN to the service provider's system running TCP/IP. The TCP/IP networking protocol can be part of your network or added as a third-party program. The leading TCP/IP program for Microsoft Windows is NetManage's Chameleon, which includes a full suite of tools for working on the Net. These tools include a Web browser, e-mail, FTP, Gopher, and others.

A growing number of Internet service providers offer dialup ISDN connections at affordable prices. Chapter 21 lists them.

Chapter 12 explains an easy approach to connecting to the Internet via ISDN. It explains using PSI's InterRamp service — a nationwide Internet service provider offering ISDN connections with an ISDN adapter card and Internet Chameleon for TCP/IP. ■

ISDN Remote Access Primer

Remote access via ISDN is based on the core concepts of internetworking and client/server computing. When you use ISDN to connect to another network you're instantly in the internetworking and client/server realm. ISDN is a network that acts as a conduit for connecting networks to networks or PCs to networks. In other words, ISDN is a cloud technology that delivers information through networks in a way that is transparent to users of networks attached to ISDN. To network users the operation of the cloud looks like a point-to-point connection even though the information can travel over different links.

Client/Server Computing

Client/server computing is the foundation for networking, which basically means one computer acts as the *host* or *server* while the other computer acts as a client. In the case of remote access, your PC acts as the client computer that connects to a server. These servers can be an online service, a BBS, a LAN, an Internet service provider, or any computer that accepts your incoming connection.

At the heart of the client/server model is the splitting of application functions between the client and the server. The World Wide Web is an example of a client/server computing system. You use a Web client program such as Netscape or Mosaic that connects to Web servers via the Internet. Information is downloaded to the client computer, which then creates the Web pages using the Web browser. The result is that computer processing is distributed, with less data required to be transmitted. ISDN's role in client/server computing is as a background conduit for connecting client and server computers. Figure 9-1 shows client/server computing in action using the Internet and ISDN.

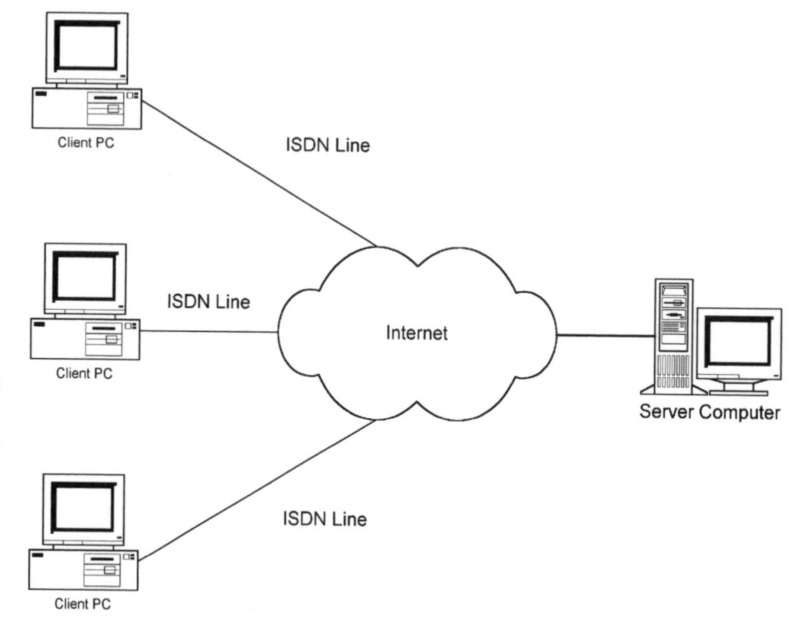

Figure 9-1: Client/server computing forms the basis of the Internet.

Networking Protocols Primer

The term protocol is used extensively when talking about networking standards. A protocol is a set of rules that specify how network communication occurs. Each network operates through protocols, with remote access connections based on both the client and server supporting the same protocol. The leading networking protocols are listed in Table 9-1. When you choose any ISDN remote access option, make sure it supports the protocols of the networks you plan to access.

Table 9-1: Leading network protocols for PC-based internetworking

Network Protocol	*Description*
TCP/IP (Transmission Control Protocol/Internet Protocol)	The networking protocol that forms the basis of the Internet. TCP/IP was developed by the Department of Defense. It's also part of the UNIX operating system.
IPX (Internet Packet Exchange)	Novell's NetWare internetworking protocols.
SPX (Sequential Packet Exchange)	
IEEE 803.2	The protocol that defines an Ethernet network at the physical layer of network signaling and cabling.
PPP (Point-to-Point Protocol)	One protocol that can be used to connect to the Internet via TCP/IP.
PPP/MP (Point-to-Point Protocol/Multilink Protocol)	A new protocol for connecting to the Internet via ISDN using both B channels.
NetBIOS	Developed by IBM and used in DOS-based networks.
NetBEUI (NetBIOS Extended User Interface)	Microsoft's implementation of NetBIOS used in Windows for Workgroups.

Internetworking

Networks are prolific in the modern computing world, and communicating across different networks (referred to as *internetworking)* is essential. The role of internetworking is to connect networks that are based on different protocols.

There are two classes of internetworking functions: bridges and routers. A bridge is a simple device that passes data from one network to another. A router is a more sophisticated device that allows data to be routed to different networks based on the packet address information associated with the data. A router is typically used to handle a variety of internetworking functions. A bridge is typically connected to a single PC or a small LAN to connect using only one network protocol at a time. However, a bridge can support multiple network protocols.

A wide area network (WAN) is a network created by connecting two or more networks that are physically isolated from one another. The connection between these networks is where ISDN comes into play. To communicate between dispersed networks you need a conduit that lets the networking protocols work with each other over long distances. Figure 9-2 shows the role of ISDN in connecting two networks via bridges.

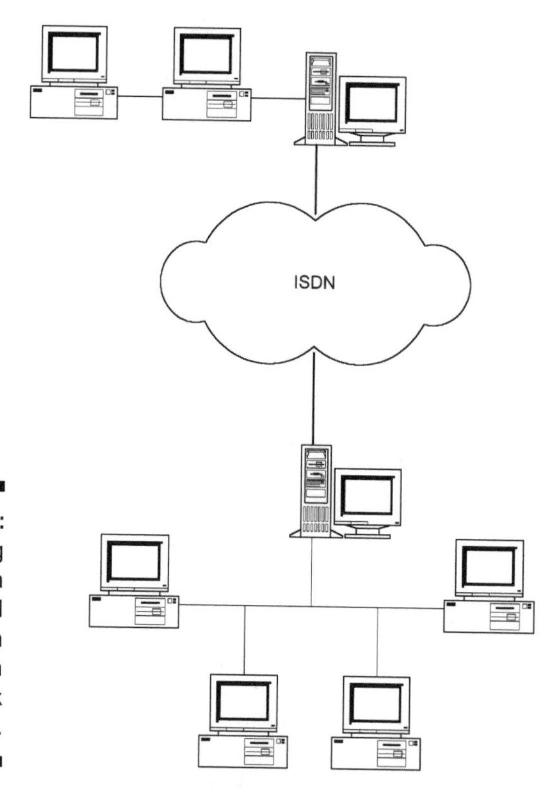

Figure 9-2: Connecting LANs via ISDN creates a wide area network (WAN).

NDIS and ODI

The network BIOS expects to see the data sent by a network adapter in a specific format. Similarly, the network BIOS sends data to the network driver in a specific format. The network adapter driver receives the information and converts it into a format the network interface card (NIC) understands. This function is handled by a software driver that acts as an interface between the network adapter card and the network operating system, called NOS for short.

The network driver interface specification (NDIS) developed by Microsoft provides a common set of rules for network adapter manufacturers and network operating system (NOS) developers for communication between the network adapter and the NOS. Most network adapters now ship with an NDIS driver. If your NOS supports NDIS, which most do, you can use any network adapter with an NDIS driver. Figure 9-3 shows the relationship of the NDIS interface and ODI interface (explained below) to the Ethernet adapter hardware and network operating system.

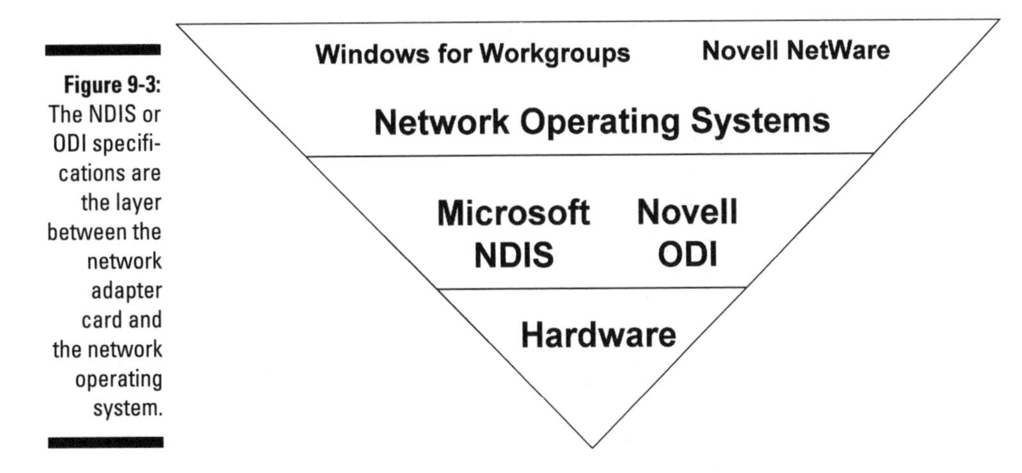

Figure 9-3: The NDIS or ODI specifications are the layer between the network adapter card and the network operating system.

Although NDIS provides compatibility between any network adapter and NOS that supports it, the primary reason NDIS was developed was to support multiprotocol stacks. These stacks enable you to run different protocols concurrently with the same network adapter. Running multiprotocol stacks enables you to use the network adapter in your computer for your LAN while also using it to access another network, such as the Internet using the TCP/IP protocol. Without the support for multiprotocol stacks provided by NDIS, you would have to unload one stack, then load another stack to access a system that uses a different protocol. This crude approach requires rebooting your computer.

A standard that serves the same purpose as Microsoft's NDIS is the Open Datalink Interface (ODI) specification developed by Novell. ODI supports multiprotocol stacks and is the standard supported in Novell's NetWare.

Ethernet-Based Remote Access

Ethernet forms the basis of most popular local area networks, such as Windows NT, Windows for Workgroups, and NetWare. It's the layer of the network residing at the physical level that involves signaling and cabling.

Ethernet provides a fast way for your PC to communicate via ISDN whether your PC is a stand-alone unit or connected to a network. Ethernet can transmit data across the local network at a rate up to 10 Mbps (megabits per second). Connecting your PC via an Ethernet-based ISDN adapter lets you take advantage of Ethernet's speed from your PC to ISDN.

You can connect your PC via Ethernet using an ISDN adapter card that emulates Ethernet, or a stand-alone bridge or router connected via an Ethernet adapter card in your PC. To use the Ethernet-based remote access option, you need a NOS running on your PC.

Adapter Cards

The Ethernet-based adapter card fits into an ISA slot in your PC. It combines an Ethernet adapter with an ISDN adapter. This option is typically used to connect a stand-alone PC to a LAN or other network via ISDN. On a stand-alone PC, you use a NOS that allows you to communicate via the Ethernet connection. The ISDN network interface card acts as a bridge to connect your PC to other networks. This is typically the least expensive Ethernet-based approach. Figure 9-4 shows the layout for using an Ethernet-based ISDN adapter card to connect to another network via ISDN.

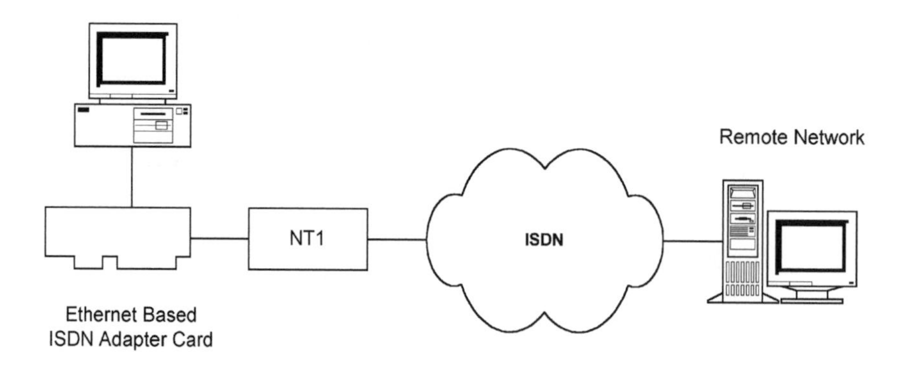

Figure 9-4: The Ethernet-based ISDN adapter card simplifies your remote access connection by consolidating the Ethernet and ISDN adapters on one board.

Chapter 11 explains working with an Ethernet-based ISDN adapter card, including installing, configuring, and using it for remote access to another network. ■

Stand-alone Single-User Bridges

The stand-alone bridge performs the same function as the adapter card but is a stand-alone device. The benefit of the stand-alone bridge is that you can connect any computer that supports Ethernet, including a PC, Macintosh, or a workstation. You connect your computer with an Ethernet card to the bridge via an Ethernet cable. The single-user bridge doesn't operate as a node on the local network. Instead, the bridge is available only to the computer connected to it. Figure 9-5 shows the layout of a typical single-user, stand-alone bridge.

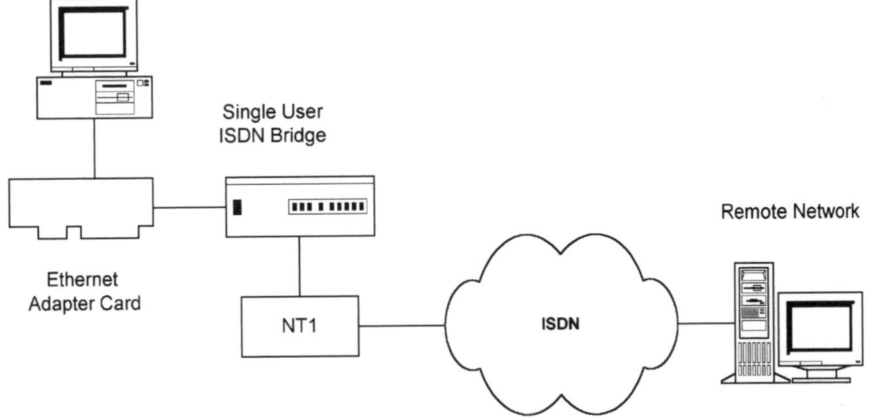

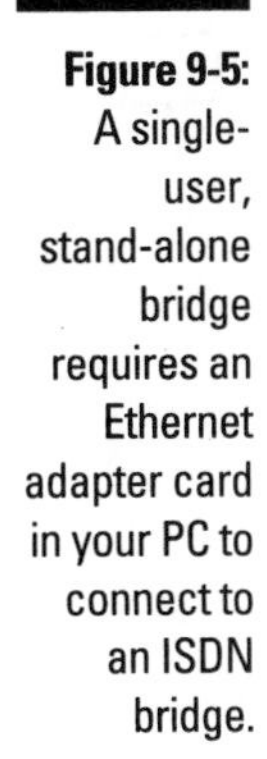

Figure 9-5: A single-user, stand-alone bridge requires an Ethernet adapter card in your PC to connect to an ISDN bridge.

Multiuser Bridges and Routers

The most powerful Ethernet-based remote access device is a multiuser bridge or router. A multiuser, stand-alone bridge or router allows you to connect a small network for remote access via an ISDN line. The device acts as an independent node on the Ethernet network, routing traffic between the local network and the remote LAN or other network through ISDN. As explained, a bridge lets multiple users connect to the same type of network via ISDN while the more sophisticated router allows data to be routed to different networks depending on the packet addresses of the data. Using a router, people on a network can send data simultaneously to different networks depending on the type of data transmitted. Figure 9-6 shows the layout of a bridge and router on a local area network.

Some ISDN CPE vendors sell devices that include both bridge and router functions. These devices allow a small office running a network to share an

ISDN connection and exchange data using a router. This stand-alone bridge and router is more expensive than a single-user bridge, but it lets you expand the number of PC, Macintosh, and UNIX machines you can connect to an ISDN line over time.

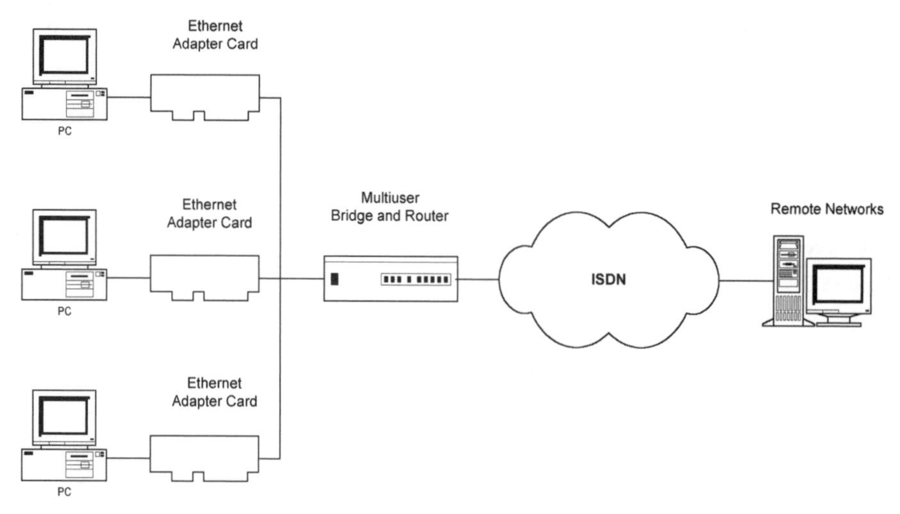

Figure 9-6: The multiuser bridge and router directs different protocol network traffic from PCs on the local network to multiple remote networks.

Serial-Based Remote Access

PCs can communicate via two forms of serial communication: asynchronous and synchronous. Asynchronous uses a start/stop bit to define each chunk of eight-bit data being sent. Synchronous data transmission sends information in larger blocks as a continuous stream using a synchronized timing method. Synchronous data transmission is the standard for digital communication, while asynchronous is the standard for analog communication. Synchronous communication is a faster method of data communication because it doesn't require the start/stop-bit overhead used in asynchronous communication. ISDN supports both forms of serial communication.

Using synchronous communications with the PPP or PPP/MP protocol lets you transfer data at the full 128 Kbps. Asynchronous communication transfers data at speeds to 115.2 Kbps via ISDN, or 57.6 Kbps for each B channel. However, the start/stop-bit method reduces the amount of data actually transferred by 20 percent. Even more of a drawback with using high-speed asynchronous communication over ISDN is that most Internet service providers and other networks don't support asynchronous communication beyond standard modem speeds.

CPE vendors market serial communication devices that include both asynchronous and synchronous communication capabilities. These devices are called *digital modems*. CPE vendors also market serial-based remote access devices that support only synchronous communication.

ISDN Digital Modems

What distinguishes digital modems from other remote access options is that they support both asynchronous and synchronous serial communication. Digital modems include a built-in asynchronous modem as well as features for converting asynchronous to synchronous communication for ISDN. They operate via your computer's serial ports and are available in two flavors: an adapter card or a stand-alone unit. The stand-alone unit connects to your PC via an RS-232 port. The digital modem solution doesn't require a NOS or network program for ISDN-based communication. It includes NT1 and TA functions but not an S/T interface port.

If the digital modem supports the PPP or PPP/MP protocol, you can connect to most Internet service providers or any computer system running TCP/IP. Connecting to a LAN using synchronous communication requires TCP/IP on both systems.

Synchronous Serial Cards

The synchronous card supports only synchronous communication, which is the standard form of communication used by ISDN. A synchronous ISDN adapter card lets you connect to any network running TCP/IP via PPP. As is the case with a digital modem, you don't need a NOS or network version of TCP/IP programs. An example of this type of product is the Internet Card used in conjunction with the PSI InterRamp Internet service, as explained in Chapter 12.

Remote Access Game Plan

Using ISDN for remote access is the most rewarding ISDN application. It's also one of the most complex in terms of options and technical details because of its inherent internetworking capabilities. As with most ISDN topics, you need to develop a strategy for remote access based on several key elements. The following sections explain the elements that you need to grasp before determining which remote access options to use.

The Zen of Remote Access

The key thing to remember in remote access is that it is internetworking. As such, your connection requires that both the client and the host communicate with the same protocols. For example, using an Ethernet-based ISDN adapter and Windows for Workgroups you can connect to a Windows for Workgroups network using its native NetBEUI protocol. To connect to the Internet via the same configuration requires a TCP/IP stack to speak the language of the Internet. However, as new protocols are supported by Windows, the capabilities of remote access will expand.

The best overall, long-term strategy for remote access is a system that offers the most versatility for support of multiple networking protocols. This versatility translates into a hardware and software system that gives you the flexibility to connect to a LAN, the Internet, another PC, or any online service.

The Great Divide: Ethernet vs. Serial

The great divide in remote access solutions is whether the remote access device uses Ethernet or serial communication to connect to ISDN. The differences in these technologies have a big impact on your remote access capabilities. Choosing the right option is your first important decision to get started with ISDN-based remote access.

In most cases going the Ethernet route is the best alternative because it lets you connect easily to both LANs and the Internet. However, the serial communication adapter card that supports only synchronous communication can offer an easier option to getting up and running with ISDN for surfing the Internet.

Digital modems are typically the least attractive ISDN remote access option. The term ISDN digital modem is a marketing term used to make the idea of connecting a PC to ISDN appear like connecting your PC to a modem. However, the digital modem route has several inherent problems, including the following:

- The high-speed asynchronous transmission touted in digital modems beyond the speeds of today's fastest modems (28.8 Kbps) isn't supported by most LANs and Internet service providers.
- The speed of 115.2 Kbps for asynchronous communication over two B channels, or 57.6 Kbps per B channel, is misleading. The start/stop bits for asynchronous communication reduce the actual data transmission rate by 20 percent to around 92 Kbps — much less than the 128 Kbps for synchronous communication.
- Most digital modems currently don't support high-speed synchronous access to common Ethernet network protocols. This means that to connect to a LAN, both client and host must use TCP/IP.
- Why buy another modem if you already have one? You can use your modem over ISDN by connecting it to an NT1 Plus device.
- Setting up a digital modem can be even more difficult than setting up an Ethernet-based ISDN remote access device.
- External digital modems occupy a serial port, which is valuable real-estate on most PCs because they usually have only two RS-232 ports.

Watch Out for the NT1

There can be only one NT1 device on your side of a BRI connection. Where the NT1 device is located is important if you plan to connect multiple devices to your ISDN line — which you'll want to do in most cases. Most vendors offer remote access devices in both U-interface and S/T-interface models. You don't want an ISDN remote access device with built-in NT1 unless it offers at least an S/T-interface port for connecting another ISDN device. Even then, it's a safer bet to get a separate NT1 or NT1 Plus device and plug your remote access device into it.

Demand PPP/MP

Until recently, connecting to the Internet via ISDN usually required proprietary implementations of the Point-to-Point Protocol (PPP). The result was that, frequently, remote access devices from different vendors didn't work with one another. Also, PPP only supports 64 Kbps data communication via one B channel. The introduction and acceptance of Point-to-Point Protocol Multilink Protocol (PPP/MP) promises to change this situation. Most CPE vendors have adopted it, and it's already incorporated in a growing number of remote access products. PPP/MP allows remote access devices to connect to the Internet using both B channels and allows interoperability between different vendors' products.

When shopping for a remote access device, make sure it supports the PPP/MP protocol. If the remote access device currently doesn't support PPP/MP, check with the vendor to find out if they plan to upgrade. If they say they don't, find another product. ■

Get into a Compressing Situation

ISDN supports up to 128 Kbps of data transmission without compression. Adding compression to ISDN communication can generate data transmission speeds up to 512 Kbps. The 512 Kbps rate means a four-fold increase over the 128 Kbps rate. This is referred to as a 4:1 compression ratio. A growing number of remote access products include built-in compression schemes. However, a compression scheme is only useful if it's supported on both ends of a connection. The good news about compression is that the Stac Hardware and Lempel Ziv Algorithm are already de facto compression standards. Additionally, a standard called Compression Control Protocol (CCP) will be implemented in the near future. It will allow two devices to determine which type of compression algorithm they both support and then carry on from there.

Let's Get Remote Accessing

You've got your grounding for remote access. The next chapter provides a survey of leading ISDN remote access devices, including Ethernet-based ISDN adapter cards, stand-alone bridges and router combo devices, and a smattering of serial communication options.

Chapter 10

ISDN Remote Access Options

In This Chapter

- Deciding which ISDN remote access solution is right for you
- Understanding the Ethernet-based ISDN adapter card option
- Looking at the single-user, stand-alone Ethernet ISDN option
- Exploring the multiuser Ethernet ISDN remote access option

Making the right choice about which remote access device to use is pivotal to getting the most from your ISDN connection. There are a growing number of available Ethernet-based ISDN remote access options, many of which offer excellent solutions. This chapter guides you through the current crop of Ethernet-based remote access devices.

The Changing Remote Access Scene

As of the writing of this book, a new generation of remote access products is emerging. These second-generation ISDN remote access devices cost less, do more, and are easier to use than their predecessors. For example, older-generation stand-alone, single-user Ethernet ISDN bridges cost more than $1000 — excluding Ethernet adapter cards. The latest Ethernet-based adapter cards can do more than stand-alone devices, don't require Ethernet network cards, and sell for as little as $499. Examples include the Combinet EVERYWARE 1000 series. The addition of Point-to-Point Protocol/Multilink Protocol and compression into second-generation remote access devices is making remote access even better.

At the time of this book's writing, a number of a new generation of remote access products that include the PPP/MP standard haven't been released. However, most vendors of remote access products support PPP/MP. Check with vendors before buying any remote access product to make sure it supports PPP/MP. Chapter 22 provides a complete listing of remote access vendors and products. ■

Ethernet-Based Remote Access

Ethernet is the foundation of most popular LANs and it provides a fast way for your PC to communicate via ISDN, whether you're using a stand-alone or networked PC. Using an Ethernet-based ISDN remote access device allows you to keep all of your existing NOS software and applications in place. Ethernet transmits data across a network at a rate up to 10 Mbps. To communicate with a network, each PC must have a network interface card. The Ethernet-based ISDN remote access adapter card acts as both an Ethernet and an ISDN adapter. A single-user Ethernet ISDN bridge connects a single computer to ISDN via an Ethernet adapter. A multiuser ISDN bridge or router connects to an Ethernet network as a node for multiple users to share access to remote sites. To use the Ethernet-based remote access option, you need a network operating system or software that can communicate with a network protocol running on your PC. A typical network connection via ISDN might involve an Ethernet-based adapter card in a PC running Windows for Workgroups. This PC connects to a remote network running Windows for Workgroups or another network with an ISDN bridge or router.

Bridging Your PC by the Board

For telecommuters or anyone else using a stand-alone PC, the Ethernet-based ISDN adapter card — which is a bridge — offers the easiest and least expensive way to connect to both LANs and the Internet. Ethernet-based ISDN adapter cards also provide an affordable option for remote access via ISDN compared to single-user stand-alone bridges. These boards combine a network interface card with an ISDN adapter card. They allow a single PC to connect to an Ethernet-based network and access all the shared resources, including applications, files, and devices available at the remote site. This approach makes the connection seamless to the person at the remote location.

You insert the adapter card into an ISA slot in your PC. Typically, these cards come in both U-interface and S/T-interface models. You use a Microsoft Windows terminal program such as the Windows Terminal program to configure the card via a COM port. The adapter card dials the ISDN telephone number of the remote LAN site. At the remote site, the LAN can use a multiuser bridge or router to receive the incoming call. If the connection is being made to another stand-alone PC, the remote PC can use an Ethernet-based ISDN remote access device. You can set up the remote access adapter to automatically make a connection to a remote site any time you locally work with an application that creates data addressed to the remote site.

Common features that should be included in any adapter card are as follows.

- Support for most popular network operating systems, including Novell's NetWare, Windows for Workgroups, Windows NT, Windows 95, and TCP/IP.
- Support for leading TCP/IP programs for working on the Internet, such as Microsoft TCP/IP or Netmanage's ChameleonNFS.
- An S/T-interface model so you can use it with other ISDN applications.
- Built-in data compression that allows file transfer at rates up to 512 Kbps over two aggregate B channels.

Where's the NT1?

When you get an adapter card remote access device with built-in NT1, at a minimum make sure it includes an S/T-interface port. The port allows you to use an additional ISDN device. However, most S/T-interface ports in Ethernet-based ISDN adapter cards require that power for the connected ISDN device is supplied by the device. For example, if you want to use an ISDN telephone, it must have its own power to work through the adapter card.

In most cases, you're better off avoiding adapter cards with built-in NT1 even if they include an S/T-interface port. The best strategy for ISDN remote access flexibility is to use a separate NT1 or NT1 Plus device. ■

The Combinet EVERYWARE 1000 Series

The Combinet EVERYWARE 1000 series of Ethernet-based ISDN adapter cards (Figure 10-1) represent the latest generation in remote access devices. They're less expensive and easier to work with than earlier remote access products. They are also the first cards to incorporate the new PPP/MP technology for Internet connections.

The EVERYWARE 1000 series of ISA bus PC adapter cards provide affordable high-speed access to Ethernet LANs and Internet service providers. These cards offer an ideal solution for the telecommuter or small business. Plugged directly into your PC's ISA bus slot, the EVERYWARE 1000 looks like an Ethernet card to your PC's software. It is compatible with popular network operating systems, including Windows, Novell NetWare, and TCP/IP.

The EVERYWARE 1000 series includes four models, as listed in Table 10-1. The models that include data compression let you transfer files at speeds up to 512 Kbps using both B channels, provided the other end of the connection also supports compression.

Figure 10-1: The EVERYWARE 1000 Ethernet-based ISDN adapter card offers an affordable remote access solution.

Chapter 11 takes you through the installation, configuration, and use of an EVERYWARE 1000 adapter card with Windows for Workgroups. ■

Table 10-1: Combinet's EVERYWARE 1000 series models

EVERYWARE 1000 Model	*Features*
PC-1030	EVERYWARE 1000 card with an S/T-interface port, but without compression or built-in NT1. $499.
PC-1040	EVERYWARE 1000 card with an S/T-interface port and built-in NT1, but without built-in compression. $699.
PC-1050	EVERYWARE 1000 card with an S/T-interface port and built-in data compression, but without built-in NT1. $599.
PC-1060	EVERYWARE 1000 card with an S/T-interface port, data compression, and built-in NT1 $899.

Digi International's DataFire

Digi International markets an Ethernet-based ISDN adapter card called DataFire. As of the writing of this book, the DataFire ISDN adapter card only ships with a built-in NT1 but no S/T-interface port. However, Digi International plans to release an S/T version. The new version of the DataFire adapter card should also support PPP/MP.

The DataFire works with leading PC network operating systems, including Windows NT, Windows for Workgroups, Novell's NetWare, and TCP/IP. It also supports data compression.

Other Ethernet-Based ISDN Adapter Cards

Other vendors in the ISDN remote access market will be introducing their own new generation of Ethernet-based ISDN adapter cards. These vendors include Intel, Network Express, Cisco, and 3Com. Look for new remote access cards from these vendors in the near future.

Chapter 22 lists the leading vendors of remote access products. Check with the vendors for more information about their Ethernet-based ISDN adapter products. ■

Single-User, Stand-alone Bridges

A single-user, stand-alone bridge connects to a computer via an Ethernet network adapter card. It is a more expensive option than the Ethernet-based ISDN adapter card. However, the key advantage of the stand-alone Ethernet bridge is that you can connect it to any PC, Macintosh, or workstation that supports Ethernet. A single-user bridge connects to a nonnetworked computer. As is the case with an Ethernet-based ISDN adapter card, using it precludes your computer from being connected to a local network. Beyond the expense of the bridge, you're also need an Ethernet network adapter card and Ethernet cabling.

You configure the stand-alone bridge with a serial cable and the Windows Terminal or other communication program. You connect the serial cable to a serial port in your PC and to a serial port at the back of the bridge device. You then use the communication program to configure the device. Once you configure your bridge, you make connections in the same way as with an Ethernet-based ISDN adapter card.

If you don't have an Ethernet network adapter card, make sure you get one that supports the same connectors as the ones on the stand-alone bridge. The three types of connectors for Ethernet networks are the DB-15 connector for 10BASE5, the BNC for 10BASE2 (BNC), and the RJ-45 for 10BASE-T.

Stand-alone, Single-User Bridge Products

A number of vendors offer stand-alone bridge products, including Combinet, ASCEND Communications, and Digi International. Single-user, stand-alone bridges represent the first generation of Ethernet-based ISDN adapters. If you're using a stand-alone PC, your best bet is an Ethernet-based ISDN adapter card. Or for the price of a single-user, stand-alone bridge, you can buy a second-generation multiuser bridge and router that support up to four users. The next section explains these multiuser devices.

The Combinet EVERYWARE 150 and EVERYWARE 160 are single-user, stand-alone bridges. Combinet offers an upgrade of these products that includes the feature sets of their new generation EVERYWARE 2000 stand-alone bridge and router series, which support multiple users. ■

Going the Multiuser Bridge and Router Route

Routers allow a single user or small office to connect one or more computers via an Ethernet network to remotely access other LANs or the Internet. These devices support PCs, Macintoshes, and workstations. Multiuser bridges and routers are a node on the network that allows computers to route their connections to external hosts on demand. Routers are more sophisticated devices than bridges because they allow a network to connect to multiple-protocol networks. Data sent by a computer on the network are identified according to their type and routed accordingly.

EVERYWARE 2000 Series

The EVERYWARE 2000 series provides single users, small businesses, and remote offices with sophisticated router capabilities. The EVERYWARE 2000 family offers the benefits of IP and IPX routing, concurrent bridging, Point-to-Point Protocol Multilink Protocol, Simple Network Management Protocol (SNMP) management, and security features. SNMP is a protocol for managing TCP/IP network traffic. The EVERYWARE 2000 security features prevent unauthorized access to the local network. The EVERYWARE 2000 series also includes compression and an Ethernet port for LAN data. The EVERYWARE 2000 is available in four configurations, as listed in Table 10-2.

Table 10-2: EVERYWARE 2000 series models

EVERYWARE 2000 Series Model	*Features*
CB-2050B	Supports up to four users. Includes an S/T-interface port and compression, but not built-in NT1. $999.
CB-2050D	Supports an unrestricted number of users. Includes an S/T-interface port and compression, but not built-in NT1. $1499.
CB-2060A	Supports up to four users. Includes an S/T-interface port, compression, and built-in NT1. $1199.
CB-2060D	Supports an unrestricted number of users. Includes an S/T-interface port, compression, and built-in NT1. $1699.

The EVERYWARE Connection Manager

The EVERYWARE Connection Manager is a Windows-based network management program for Combinet's network access products. This software simplifies setting up a LAN to manage incoming and outgoing ISDN calls in conjunction with Combinet EVERYWARE 2000 routers. It provides centralized dial-in user authentication and call management facilities, as well as configuration, performance monitoring, and accounting features to help manage the central LAN. Once installed and enabled, the Connection Manager handles call processing, security checking, distribution, and accounting for both local and remote ISDN access devices. It also includes such features as automatic callback.

Other Multiuser Stand-alone Bridges and Routers

Digi International's soon-to-be-released Retoura Touch Router represents another second generation of Ethernet-based ISDN routers. Other vendors of multiuser bridges and routers include ASCEND Communications, Gandalf, 3Com, Cisco, and Microcom. Chapter 22 provides a more extensive listing of multiuser bridge and router vendors and products.

On to Hands-On Remote Access

You have a number of ISDN remote access options available, but one of the best is the Ethernet-based ISDN adapter card. The next chapter takes you along during installing, configuring, and using the Combinet EVERYWARE 1000. This Ethernet-based ISDN adapter card represents the new generation of ISDN remote access equipment.

"OK, TECHNICALLY THIS SHOULD WORK. JUDY, TYPE THE WORD, 'GOODYEAR' ALL CAPS, BOLD FACE, AT 700 POINT TYPE SIZE."

Chapter 11

Using the Combinet EVERYWARE 1000

In This Chapter

- Installing the EVERYWARE 1000 adapter card and network adapter software
- Configuring the EVERYWARE 1000 software for Windows for Workgroups
- Connecting to a Windows for Workgroups network using the EVERYWARE 1000

Combinet's EVERYWARE 1000 ISDN adapter card represents the latest in Ethernet-based ISDN remote access devices. It provides an economical, easy-to-install, and high performance remote access option that is ideal for the telecommuter or small business. This chapter takes you on a walk-through of setting up and using the EVERYWARE 1000 with Windows for Workgroups.

Before Installing the EVERYWARE 1000

Before installing the EVERYWARE 1000 adapter card, you need to make sure that the card's default I/O address, IRQ level, COM port selection, and interrupt settings don't conflict with another board in your PC that can't be changed. In most cases the adapter card's preconfigured settings don't conflict with other devices. However, if you need to make changes, the EVERYWARE 1000 provides extensive alternate settings. You make these changes using the 10-position DIP switch on the slot plate of the adapter card. This handy location for the DIP switch gives you access to the switch even with the card installed. For the default settings, all 10 DIP switch settings are OFF (in the up position). Table 11-1 lists the default settings for the EVERYWARE 1000 adapter card.

If you're unsure whether the default settings of the EVERYWARE 1000 card will clash with any other settings, you can check the status of memory addresses and IRQ settings on your PC with the MSD program. The MSD (Microsoft System Diagnostics) program ships with DOS 6.0 or later and is a handy tool for checking what is happening inside your computer. Use this program to check if the default settings of the EVERYWARE 1000 board are available or used by another device. To use the MSD program, at the DOS prompt type MSD.

Table 11-1: Default settings for the EVERYWARE 1000 adapter card

Setting	*Default*
Input/Output (I/O) Base Address	0300-031F
IRQ Level	15
Communications Port	COM4
Communications Port IRQ Level	5

Installing the EVERYWARE 1000 Adapter Card

You can install the EVERYWARE 1000 in any available 16-bit ISA expansion slot in your PC. Once you install the card, you connect the cables to the NT1 device or to the U-interface jack for your ISDN line, depending on the EVERYWARE 1000 model you're installing.

If you want to use another ISDN device, such as an ISDN telephone, on the same ISDN line as the EVERYWARE 1000, you must terminate one of the two devices. In general, the device that is farthest from the NT1 is the one to terminate. The default setting for the EVERYWARE 1000 is the ON position for termination. You change the termination by removing a set of two jumpers located at the top of the adapter card. Don't change the setting unless you actually plug in the other ISDN device and set the termination on that device.

If you already have an Ethernet card in your PC, remove it before installing the EVERYWARE 1000 card. Your PC can support only one Ethernet card. ■

To install the adapter card without another ISDN device, do the following.

1. If you can use the default settings, don't change the DIP switch settings. If you need to change any card settings, make the changes using the 10 DIP switch settings.
2. Install the adapter card in any available 16-bit ISA bus expansion slot of your PC.
3. Connect the cable. If you're using the 1040 or 1060 model, connect the unit directly to the ISDN wall jack. If you're using the 1030 or 1050, connect it to your NT1 or NT1 plus device's S/T-interface port.

Installing the Network Adapter Software

After installing the adapter card, the next step is installing the EVERYWARE 1000 network adapter software from DOS. This network adapter software is loaded every time you boot your PC and requires about 2.5MB of disk space.

Before you install the EVERYWARE 1000 adapter card's software, check the status of your conventional memory, which is your PC's memory up to 640K. The driver for the EVERYWARE 1000 takes about 49K of RAM space, so if you have a lot of programs crowded into conventional memory, you may have a problem. If you need to make room in your conventional memory, you can use the DOS 6 MemMaker command or a third-party memory manager before installing the network adapter software. MemMaker modifies your CONFIG.SYS and AUTOEXEC.BAT files so that your device drivers and other memory-resident programs use less conventional memory.

The network adapter software creates a PCBRI directory for the EVERYWARE 1000 files. The program also modifies your CONFIG.SYS and the AUTOEXEC.BAT files so the EVERYWARE 1000 adapter can work properly. The EVERYWARE 1000 installation program adds the following statement in your CONFIG.SYS file, which must be the first line:

DEVICE=C:\PCBRI\swl.exe-rl C:\PCBRI\sbp26.dms

This statement instructs DOS to load software into the adapter card every time you boot your PC.

The following steps explain how to install the network adapter software.

1. At the DOS prompt, insert the Combinet Adapter Card disk in your floppy drive and change to that floppy drive (A: or B:).
2. Type *install* and press Enter. The installation program's welcome screen appears.

3. Press Enter. A dialog box appears for entering the destination drive and directory of your files. By default, the destination path is C:\PCBRI, but you can enter a different drive and directory.
4. Press Enter if you want to use the default path, or enter a new path and press Enter. A new message box appears.
5. Use your arrow keys to highlight the software option you need. These options specify the type of telephone company switch used for your BRI connection. Table 11-2 describes these switch options. After highlighting the option you want, press the spacebar to select it.
6. Press Enter. The installation program displays a status slider as it progresses with the installation.

Table 11-2: Switch setting options for the EVERYWARE 1000

Option	*Description*
5ESS Software Installation	AT&T 5ESS switch running Custom software
DMS, NI-1 Software Installation	AT&T 5ESS switch running National ISDN-1 (NI-1) software, or a DMS switch running proprietary software
NET3 Software Installation	Europe ISDN
1TR6 Software Installation	1TR6 (Germany)
VN3 Software Installation	VN (France)
INS Software Installation	DMS switch running NI-1

Installing the Microsoft Windows Network Driver

The EVERYWARE 1000 adapter card emulates an Ethernet adapter using Open Datalink Interface (ODI) and Network Driver Interface Specification (NDIS) software drivers. The ODI version is compatible with Novell NetWare 3.X and 4.X environments. The NDIS version 2.01 driver is compatible with DOS, Windows 3.1, and Windows for Workgroups 3.10 and 3.11.

Before you install the NDIS software for Windows for Workgroups, you should have already installed the EVERYWARE 1000 adapter card software from DOS.

If you don't have Windows for Workgroups set on your system as a network, make sure you have the Windows for Workgroups disks. You'll need to install Windows for Workgroups networking software as part of the EVERYWARE 1000 Windows networking driver installation.

Here's how to install the NDIS2 driver for Windows for Workgroups.

1. Start Windows. If you previously had an Ethernet card in your computer, two warning boxes appear. Choose OK for each message box. If you haven't installed an Ethernet card on your system, the warning dialog boxes don't appear.
2. Open the Network group in Program Manager, then double-click the Network Setup icon. The Network Setup dialog box appears, as shown in Figure 11-1.
3. Click the Drivers button. The Network Drivers dialog box appears. If any network adapter card driver is listed in the Network Drivers list, highlight the driver and click the Remove button. Click the Yes button to confirm the removal of the driver.
4. Click the Add Adapter button in the Network Drivers dialog box. The Add Adapter dialog box appears.
5. Highlight the Unlisted or Update Network Adapter option in the list of network adapters and choose OK. The Install Driver dialog box appears.
6. Insert the Combinet Drivers disk into your floppy drive and choose OK. The Unlisted or Updated Network Adapter dialog box appears with the newly installed PC BRI LAN ADAPTER CARD driver name in it.
7. Choose OK from the Unlisted or Updated Network Adapter dialog box. The Network Drivers dialog box appears.
8. Highlight the PC BRI LAN Adapter Card (NDIS2/NDIS3) entry in the Network Drivers list and click the Setup button.
9. Highlight the REAL Mode NDIS entry in the Driver Type drop-down list and choose OK. The PC BRI LAN Adapter Card dialog box appears.
10. Choose the IRQ setting for your adapter card from the Interrupt (IRQ) drop-down list. After choosing the IRQ setting, choose OK. The Network Drivers dialog box appears with your newly-added PC Adapter Card NDIS driver, as shown in Figure 11-2.
11. Choose Close from the Network Drivers dialog box. The Network Setup dialog box appears listing your installed PC Adapter Card NDIS2 driver.

12. Choose OK in the Network Setup dialog box. A Windows Setup dialog box appears, asking if you want to replace files for the Enhanced Protocol Manager currently installed. Choose No. A Windows Setup information dialog box appears, informing you that modifications were made to your SYSTEM.INI and PROTOCOL.INI files. Choose OK. A message dialog box appears prompting you to restart you computer.

13. Choose Restart Computer to restart your computer and enable the PC BRI LAN Adapter Card Installation.

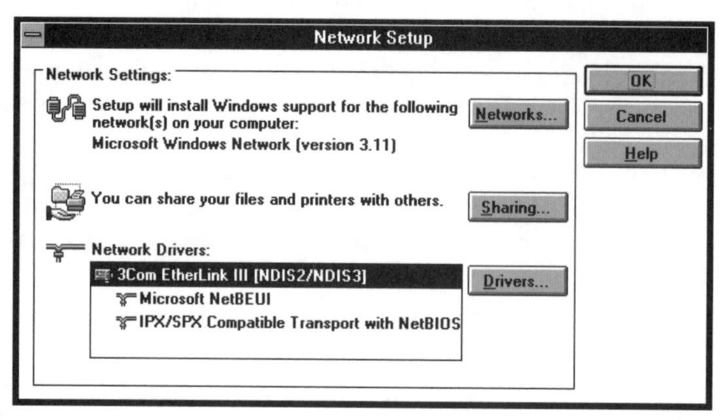

Figure 11-1: The Network Setup dialog box is used to install the NDIS driver for Windows for Workgroups.

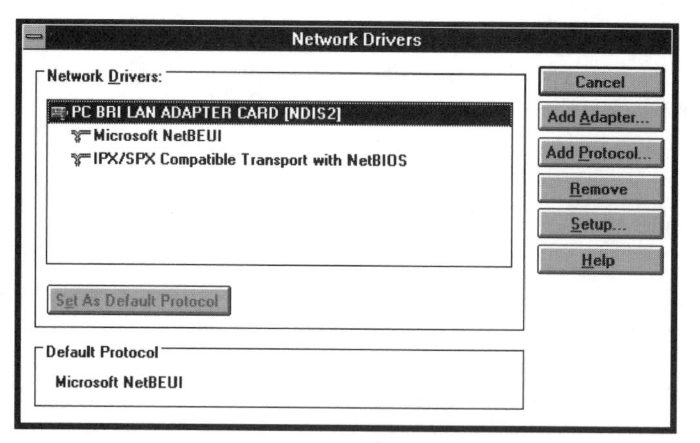

Figure 11-2: The Network Drivers dialog box with the newly-added PC BRI LAN ADAPTER CARD (NDIS2) driver.

Configuring the EVERYWARE 1000

Once you've installed the network driver, you need to perform a software configuration for the EVERYWARE 1000. The configuration of the adapter card requires using the Terminal program in Windows or a third-party communication program. You enter commands to configure the EVERYWARE 1000 using simple commands.

Setting Up Your COM Port

Before you can use Windows Terminal (or any other communications program) to communicate with your EVERYWARE 1000 adapter card, you must configure the communication port from Windows. The following is a step-by-step procedure for configuring the COM4 port in Windows, which is the default setting for the EVERYWARE 1000 adapter card.

1. In the Windows Control Panel, open the Ports icon. The Ports dialog box appears (Figure 11-3).
2. Select COM4 and choose the Setting button. The settings dialog box for COM4 appears.
3. Click the Advanced button. The Advanced Settings for the COM4 dialog box appears, as shown in Figure 11-4.
4. Choose 02E8 from the drop-down list in the Base I/O Port Address setting.
5. Choose 5 from the drop-down list in the Interrupt Request Line (IRQ) setting, then choose OK.
6. Choose OK. A System Setting Change dialog box appears. Choose Restart Now. Windows restarts.
7. Open the Windows Accessories group and double-click the Terminal icon. The Terminal window appears.
8. Choose Settings | Communications to display the Communications dialog box (Figure 11-5). Make sure the COM4 option is highlighted in the Connector list, then choose OK. You're now ready to configure the EVERYWARE 1000.

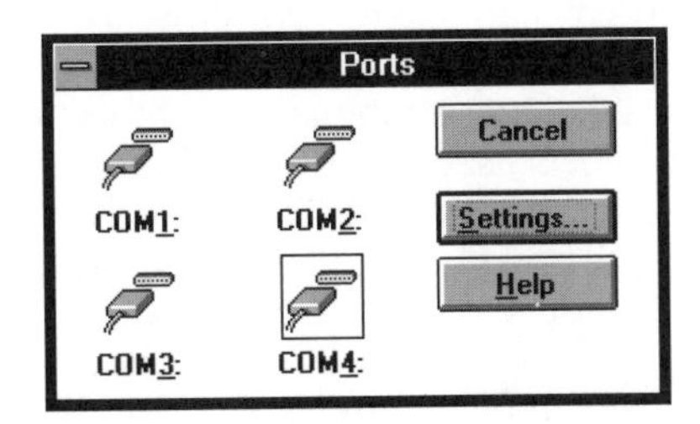

Figure 11-3: The Ports dialog box lets you configure Microsoft Windows for the COM4 port used by the EVERYWARE 1000.

Advanced Settings for COM4:
Base I/O Port Address: 02E8
Interrupt Request Line (IRQ): 5
OK
Cancel
Help

Figure 11-4: The Advanced Settings for COM4 dialog box lets you enter the Base I/O Port Address and Interrupt Request Line for COM4.

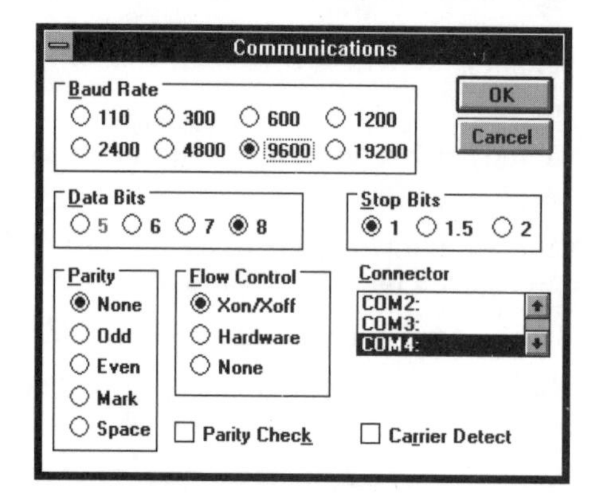

Figure 11-5: Specify the COM4 setting in the Communications dialog box's Connector list.

Basic Configuration of the EVERYWARE 1000

Once you've set up the port so you can use the Windows Terminal program, you can configure the EVERYWARE 1000. The basic configuration of the EVERYWARE 1000 tells it your ISDN line configuration information, such as the directory number and any SPIDs used by your ISDN line. This configuration changes the firmware of the adapter card. The EVERYWARE 1000 uses these settings as its defaults until you change them. Beyond the basic configuration commands, there is a collection of commands to configure and operate the EVERYWARE 1000 adapter. The following instructions explain how to configure the EVERYWARE 1000 to work with your ISDN line using the Windows Terminal program.

1. Open the Terminal window.
2. Enter SE DI *telephone number*, for example: SE DI 14155551212. This command tells the adapter card your ISDN line directory number.
3. If the telephone company assigned you a single SPID for your line, enter SE SPID *SPID number*, for example: SE SPID 50855512120000. If your telephone company assigned you two SPIDs, enter them separately for each channel. Enter SE L1 SPID *SPID number* and SE L2 SPID *SPID number.*
4. Enter SE SPE AU. This command instructs the EVERYWARE 1000 to attempt the first call at 64 Kbps. If the call is unsuccessful, the second call attempt is sent at 56 Kbps. As you recall, long distance ISDN calls are typically at 56 Kbps.
5. Enter SH to check your configuration entries. A list of configuration settings appears in the Terminal window, as shown in Figure 11-6. Press Enter to scroll down the list to view all the settings options.

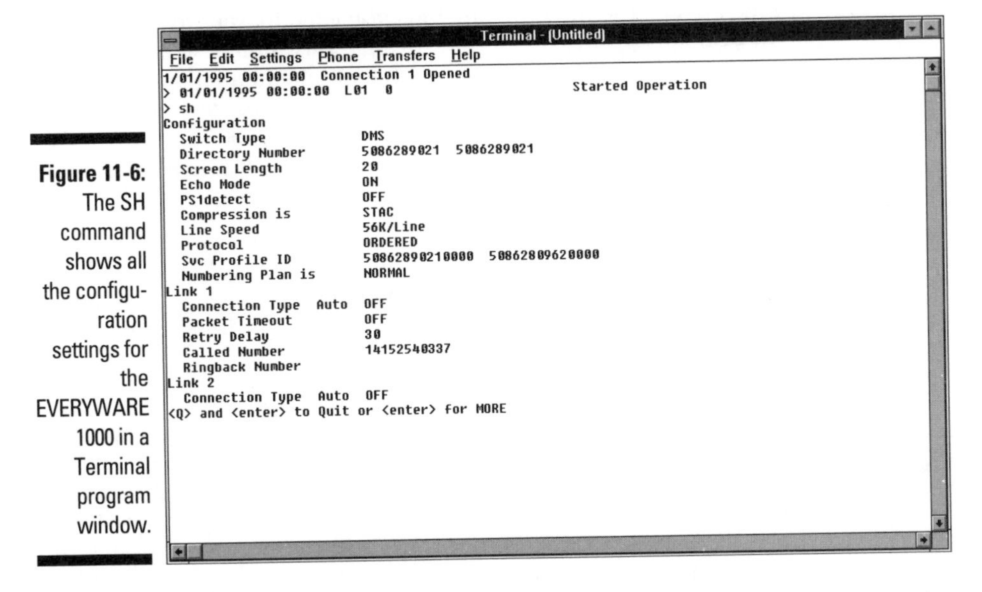

Figure 11-6: The SH command shows all the configuration settings for the EVERYWARE 1000 in a Terminal program window.

Basic EVERYWARE 1000 Commands

You can enter commands via the Windows Terminal program to configure and operate the EVERYWARE 1000. For example, you can configure the EVERYWARE 1000 to automatically make a connection any time you work with an application that sends data to the remote system.

Commands for configuring the EVERYWARE 1000 begin with SE, which is short for Set. Commands to operate the EVERYWARE 1000, such as for making calls, typically include specifications for each B channel. The following lists several common commands for configuring and operating the EVERYWARE 1000.

- The CA (call) command initiates a call on a specified channel to a specified telephone number. For example, CA 14085551212 calls the single number for one B channel. To initiate a call using both B channels, you enter CA L1 *phone number* and CA L2 *phone number.* The L1 and L2 refers to link 1 and link 2, which translate to B channels 1 and 2.
- The DI (disconnect) command terminates the call on the specified channel. For example, DI L2 disconnects the call on channel 2.
- The SE NU (set number) command specifies the phone numbers that each link will call when you use the call command or for on-demand calls. For example, entering SE NU 14085551212 instructs the EVERYWARE 1000 to call that number any time you execute the call command. Using this command saves you from entering the number each time. You can specify a number for each link. The telephone number specified with the set number command also specifies the number for on-demand dialing. On-demand dialing means that the EVERYWARE 1000 automatically makes a connection to the remote site any time data traffic is generated on your PC for a remote network. To use on-demand dialing, you must use the set auto on and the set time commands.
- The SE AU ON (set auto on) command tells the EVERYWARE 1000 to automatically connect to the phone number specified using the set number command any time you work with an application that creates data addressed to the remote site.
- The SE TI (set timeout) command defines the amount of time in seconds before the ISDN disconnects if there is no Ethernet activity. For example, to set the ISDN line to disconnect after five minutes (300 seconds) of no Ethernet activity, enter SE <1 TI 300, which sets the limit for link 1. Do the same for link 2 by entering SE L2 TI 300. The range in seconds is 1 to 32,767. Use this command if you use the SEt AUto ON command to make sure you don't stay connected after doing what you need to do on the remote site. Otherwise, you could have a whopping ISDN telephone bill.
- The SH (show) command displays the adapter card's configuration and the status of both links. The SHow COnfig command displays a subset of the current configuration settings without the line status information.
- The SE CALLI (set calling receive) command specifies the phone numbers that your adapter card will accept for incoming calls. For example, enter SE CALLI 15085551212. For multiple set calling receive numbers enter the link (channel) number before the number; for example, SE CALLI L2 15085551212. This command works in conjunction the SEt CALLEr ID command, which you turn on by entering SE CALLE ID ON.

The current version of EVERYWARE 1000 software doesn't let you specify multiple sites for automatic connections. For example, you can't have the EVERYWARE 1000 connect automatically to a Windows for Workgroups remote site, disconnect, then automatically connect to an Internet service provider when you execute a TCP/IP program. However, a version of the software available in the near future from Combinet will support making multiple connections automatically using numbers you specify. ■

The EVERYWARE 1000 Card's Pretty Lights

All the EVERYWARE 1000 adapter cards include status lights on the slot plate. The EVERYWARE 1000 series models 1040 and 1060, which include a built-in NT1 function, include three LEDs to provide status information. The EVERYWARE 1030 and 1050, which don't include an NT1 function, include two LEDs. Unfortunately, these lights are at the back of your PC, so viewing them is at best a strain. Table 11-3 describes the functions of these LEDs for all the cards.

Table 11-3: LEDs on EVERYWARE 1000 adapter cards

LED	*Description*
Channel 1	For the 1040 and 1060, a blinking green LED indicates that a call is being connected on B channel number 1. When the call is established, the LED remains lit without blinking. For the 1030 and 1050, a green LED indicates that the Channel 1 line is connected and packets can be sent and received.
Channel 2	For the 1040 and 1060, a blinking green LED indicates that a call is being connected on B channel number 2. Once the call is established, the LED remains lit without blinking. For the 1030 and 1050, a green LED indicates that the Channel 2 line is connected and packets can be sent and received.
NT1	For the 1040 and 1060, a steady green LED indicates that the internal NT1 and the telephone switch are synchronized. A fast blink indicates that the card is attempting to synchronize with the telephone switch. A slow blink indicates the card is attempting to synchronize with ISDN terminal devices. For the 1030 and 1050, there is no NT1 light because these models don't include a built-in NT1 device.

Connecting to a Windows for Workgroups Remote Site

To make a connection to another Windows for Workgroups network via the EVERYWARE 1000, you need the telephone numbers to call and the workgroup name. The workgroup name lets you connect as a user who is part of that group for sharing files and devices. The functions of sharing drives and printers in Windows for Workgroups must also be set up at the remote site for you to share directories or drives. The remote PC or network to which you connect must have an ISDN remote access device that supports the NDIS driver. Typically, this device is an ISDN bridge or a router device. You can also connect to another stand-alone PC in the same way as you do for connecting to a network. Keep in mind that once the EVERYWARE 1000 makes the connection, you're working with Windows for Workgroups as the network between your system and the remote site.

1. Open the Network icon in Control Panel. The Microsoft Windows Network dialog box appears, as shown in Figure 11-7.
2. Enter the name of the workgroup at the remote site in the Workgroup field. You only need to do this once. Click OK. A dialog box appears, prompting you to restart your computer. Click the Restart Computer button.
3. Open the Terminal program in the Accessories group.
4. Enter the Call command followed by the telephone number of the remote system. For example, enter CA 14155551212 for a single link, then press Enter. To connect two channels, enter the Call command with the Link parameter before each telephone number. For example, CA L1 14145551212 and CA L2 14155551212. The connection is made to the remote network.
5. In File Manager, choose Disk | Connect Network Drive or click Connect Network Drive to connect to a network drive at the remote workgroup. The connection is made and a drive icon appears on File Manager's Drive bar.
6. To use an application on the remote site, navigate to the shared directory with the program file, then double-click the program file. The program is executed and appears on your screen.
7. To download or upload a file, use File Manager's copy or move command, or drag file icons between windows.

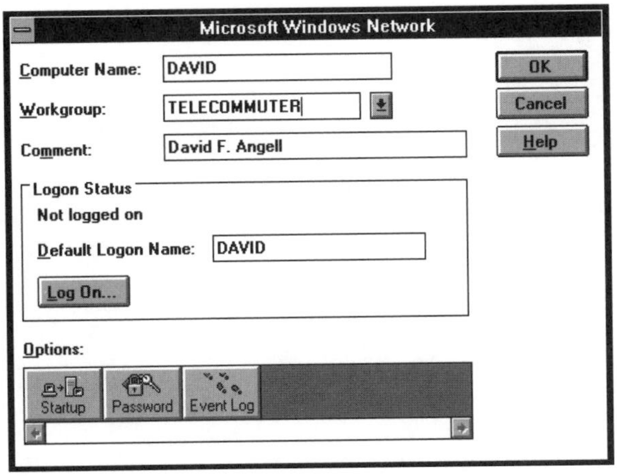

Figure 11-7: The Microsoft Windows Network dialog box is where you must add the workgroup name for a remote system.

Connecting to the Internet

Connecting to the Internet using the EVERYWARE 1000 is similar to connecting to a Windows for Workgroups network. However, you need to run a network version of a TCP/IP stack, such as the Microsoft TCP/IP stack or ChameleonNFS for Windows. You set up the EVERYWARE 1000 and initiate a call the same way you do for Windows for Workgroups. You can have the EVERYWARE 1000 automatically connect to your Internet service provider any time you execute a TCP/IP application, such as FTP or a Web browser. You can also manually make the call using the Terminal program.

On to an Easy Internet Access Option

If all you ever want to do is connect to the Internet via ISDN as cheaply as possible, the next chapter shows you the way. It explains working with a simple serial-based ISDN adapter card, Internet Chameleon, and PSI's InterRamp service.

Chapter 12

Going the Easy Internet Access Route

In This Chapter

- Connecting to the Internet with PSI's InterRamp package
- Installing and setting up the ISDN*tek Internet Card and Internet Chameleon

One of the easiest ways to connect to the Internet with ISDN is to use The Internet Card from ISDN*tek and the PSI InterRamp membership kit. This chapter gets you up and running using this inexpensive ISDN-based Internet access solution.

The Pieces of the PSI Internet Access Package

PSI is a leading Internet service provider and early pioneer in affordable ISDN access. The InterRamp membership kit includes an Internet access account and the Internet Chameleon software. To complete the InterRamp package, you also need to purchase the ISDN*tek Internet Card. The following sections explain the three parts of the InterRamp package.

The ISDN*tek Internet Card

The Internet Card provides a relatively low-cost serial solution via synchronous communication for surfing on the Internet. It connects to your PC via an ISA expansion slot and to the ISDN network via an S/T-interface. You need an NT1 or NT1 Plus device to use the ISDN*tek card. ISDN*tek's Internet Card lets you communicate with any Internet service provider as well as any LAN supporting TCP/IP. While the current Internet Card only supports a single B channel for data communication, a new version of the card will support PPP/MP (Point-to-Point Protocol Multilink Protocol). This will allow you to use both B channels on the Internet.

Buy the Internet Card from PSI. It costs $299 if you buy it from them and $395 if you buy it directly from ISDN*tek. However, you can buy an NT1 from ISDN*tek for $160, which is a good price. The NT1, covered in Chapter 6, is made by Alpha Telecom. ■

What makes installing the ISDN*tek Internet Card easy is that it supports the WinISDN Application Program Interface. WinISDN provides Microsoft Windows-based PCs with a standard interface between a serial card and ISDN-ready TCP/IP software. The result is that installing the TCP/IP software to work with serial communication boards is easier. WinISDN was developed by ISDN*tek and NetManage as an open system for board vendors to support as a way to connect software to boards. Expect to see other serial communication adapter boards support WinISDN.

Internet Chameleon

NetManage's Internet Chameleon is included in the PSI InterRamp membership kit. This is the non-network version of NetManage's popular Chameleon-NFS for Windows package. Internet Chameleon includes the TCP/IP stack for Windows and a suite of TCP/IP tools for working and playing on the Net. These Windows tools include FTP, e-mail, telnet, Gopher, a World Wide Web browser, and more.

PSI's InterRamp

PSI's InterRamp service is a nationwide system that is available via local telephone numbers. PSI provides Internet gateways or routers at Points of Presence, called POPs in phone network lingo. By calling a local POP number, you're connected to the Internet without long distance calling charges. Most metropolitan areas have local POP numbers for connecting to PSI.

Chapter 21 lists PSI's POP numbers for ISDN service. ■

The InterRamp membership kit costs $99 for the first month, which includes unlimited connection time. After the first month the price for Internet service is $29 per month for 29 hours of connect time. This price doesn't include telephone company usage charges for calls to the POP numbers.

Getting Started

Once you've assembled the InterRamp membership kit, the Internet Card, and the NT1 or NT1 Plus device, you're ready to get started. The process for connecting to the Internet via the PSI InterRamp includes the following steps.

1. Configure your ISDN line for the Internet Card and set up your NT1 or NT1 Plus device. Setting up an NT1 or NT1 Plus device is explained in Part II.
2. Install and configure the ISDN*tek Internet Card in your PC.
3. Install and configure the Internet Chameleon program.

Configuring Your BRI Line

Like most ISDN devices, the ISDN*tek card works with the leading telephone company switches, including National ISDN-1 (NI-1), AT&T 5ESS (Custom), or the NT DMS100. You can configure your channels for only data or alternate voice/data. The Internet Card supports one, two, or no SPIDs.

Setting Up the ISDN*tek Internet Card

To make installation of the ISDN*tek card easier, it ships with a Windows utility program called ISDNTEST, which is a setup and configuration program. Make sure you install and run the setup program before you install the card. To install the program, do the following.

1. From Program Manager choose File | Run, enter *A:\SETUP*, and click OK to install the setup program on your hard disk. The setup program installs the ISDN*tek files on your system and creates an ISDN Setup group in Program Manager.
2. Open the ISDN Setup group and click the Test ISDN Hardware icon. The ISDN*tek PC Board Setup window appears, as shown in Figure 12-1. The settings in the Board Settings group are the default memory address and IRQ settings for the ISDN*tek board. For now, ignore these settings.
3. Enter your SPID numbers, if your telephone company assigned any, in the SPID boxes located in the ISDN Line Info group.

4. Enter a telephone number in the Test # field to see if your ISDN line is working — after you install your card. This is an optional setting. Use a telephone number at your location if you can, so you can hear it ring. Choose the options for the type of test call you want at the bottom. In most cases, use the default Voice setting.

5. Click the Set button to save changes to the ISDN.INI file. At this point you must configure your ISDN*tek Internet Card and install it, as explained in the following section.

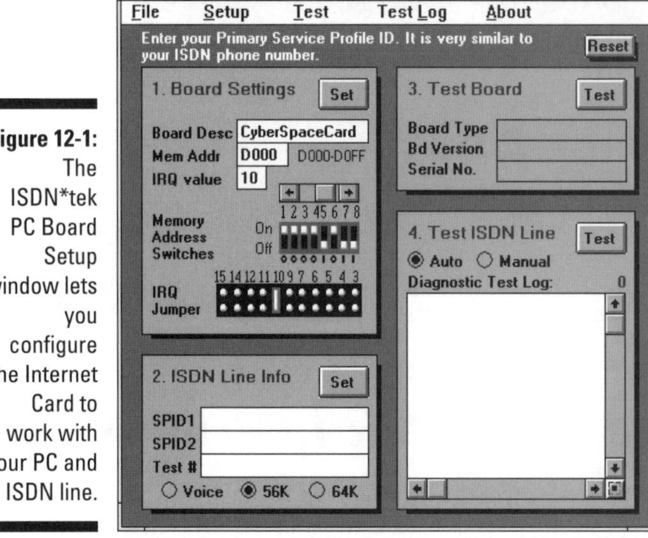

Figure 12-1: The ISDN*tek PC Board Setup window lets you configure the Internet Card to work with your PC and ISDN line.

Configuring the Internet Card for Your PC

The two settings you can change on the ISDN*tek board are the memory address and the IRQ used by the board. The default memory address for the card is D000-D0FF and the default IRQ setting is IRQ 10. The memory address specifies the location in RAM used by the ISDN*tek card. The IRQ setting specifies the interrupt request line used by the PC to get the attention of the CPU when the ISDN*tek card is ready to send or receive data.

If you know that these settings don't clash with any other devices connected to your PC, install the Internet Card in your PC and continue the installation process, as explained in the next section.

If you're unsure if the default settings of the Internet Card clash with other settings, you can check the status of memory addresses and IRQ settings on your PC with the MSD program. The MSD (Microsoft System Diagnostics) program ships with DOS 6.0 and later, and is handy for checking what is happening inside your computer, including memory allocation and IRQ settings. Use this program to check if the default settings of the ISDN*tek board are available or already are taken. To use the MSD program, type MSD at the DOS prompt.

The eight-position DIP switch located on the card lets you choose a memory address setting from a variety of choices. You can choose any available memory address and see its switch settings by entering the value in the Mem Addr field in the Board Settings group. The illustrated switch changes to show you the correct settings. You can also scroll through the memory address options using the slider just above the Memory Address Switches settings in the Board Settings group. You change these setting by moving the switches up or down depending on the configuration for the setting. Moving a switch up turns it on, while moving it down turns it off.

To change the default IRQ settings for the ISDN*tek board, you move the jumper to the IRQ you want. You can set your ISDN*tek board for IRQ 3, 4, 5, 6, 7, 9, 10, 11, 12, 14, or 15.

If you experience a problem installing your ISDN*tek Internet Card, you may need to change a BIOS setting for your PC using the PC's CMOS setup program. To activate the CMOS program on many PCs, you press Delete while memory is cycling during bootup. Your CMOS program appears. Choose the advanced CMOS setup option. Look for the Adapter Shadow ROM for the IRQ you're using and disable it.

Completing Installation of the Internet Card

After you've made any desired changes using the switches on your board, you're ready to install the board in your PC and complete installation of the ISDN*tek card. To complete this installation, do the following.

1. Install the Internet Card in any available 16-bit ISA bus slot in your PC.
2. If you changed the Memory Address switch setting on your card, you must change the settings in the Board Settings group to match, if you didn't already do so when you checked for memory addresses.
3. If you changed the IRQ jumper setting on the Internet Card, change the IRQ value setting in the Board Settings group. To change the IRQ setting, point to the value in the IRQ jumper setting and click.
4. Click the Use button in the Board Settings group. The settings are saved to the ISDN.INI file, then the Install program searches your CONFIG.SYS and SYSTEM.INI files, making recommendations regarding memory conflicts.

5. Click the Test button in the Test Board group of the Install program window. The Install program reads the information from the card and displays it in three fields: Board Type, Bd Version, and Serial No. If the fields can't be read, there is a problem with the board settings.
6. Plug the RJ-45 cable into the NT1 S/T-interface port and the Internet Card.
7. Click the Test button in the Test ISDN Line group. If you have a good line connection, several response messages appear in the Diagnostic Test Log in the Test ISDN Line group. The telephone number you entered in the ISDN Line Info group should ring.
8. Choose Setup | Save Settings to save your settings, then choose File | Exit to close the ISDN*tek PC Board Setup window.

Installing Internet Chameleon

Once you've installed the Internet Card, the next step is to install Internet Chameleon. But before you do so, make sure you have your PSI InterRamp User's Guide. The guide includes information that you need to set up Chameleon, including your account ID and password, as well as IP addresses for PSI servers. It also includes instructions for setting up Internet Chameleon to work for PSI InterRamp. The Internet Chameleon program also ships with its own documentation. Here's how to install Internet Chameleon.

1. Insert the Internet Chameleon disk in drive A:.
2. Choose File | Run in Program Manager.
3. Enter *A:\SETUP* and choose OK. The setup program loads the Internet Chameleon files onto your system.
4. At the prompt, insert the Internet NEWT disk and choose the Continue button. The setup program completes the installation of the Internet Chameleon program.

Configuring Chameleon

After you install Internet Chameleon on your system, you configure it to work with PSI. The following steps explain how to set-up Chameleon for PSI InterRamp.

1. Double-click the Custom icon in the Internet Chameleon group. The Internet Chameleon group window appears, as shown in Figure 12-2.
2. Choose File | Open. The Open Configuration File dialog box appears.

3. Select INTERAMP.CFG for PSI from the File Name list and click OK.
4. Choose Setup | Call Type, select B64K, and click OK.
5. Choose Set | Dial and enter the ISDN telephone number for accessing PSI, which is listed in your InterRamp User's Guide.
6. Choose Set | Hardware. The Hardware dialog box appears.
7. Choose the ISDN*tek option in the Vendor field. The settings for your board appear in the other boxes. These are the default settings; if you changed the IRQ or memory setting, change the corresponding default.
8. Enter your SPID numbers and click OK.
9. Choose Setup | Login and enter your username and password from the attached card on your InterRamp User's Guide.
10. Choose File | Exit. Chameleon prompts you to save the changes to the Chameleon configuration file.

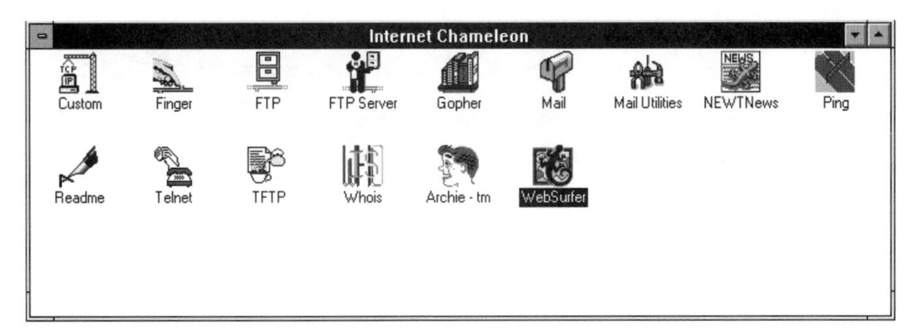

Figure 12-2: The Internet Chameleon group window displays icons for all of Chameleon's TCP/IP tools.

Setting Up E-mail

To use e-mail you configure the Chameleon E-mail tool, which is a straightforward process. You set up your own mailbox and specify the mail server at the PSI server. This information is provided in the InterRamp documentation. To set up e-mail for your PSI account, do the following.

1. Click the Mail icon and log in as Postmaster. You don't need a password.
2. Choose Services | Mailboxes, enter a name for the mailbox, and click Add. A dialog box with the name you chose for the mailbox appears in the title bar.

3. Add the name again, a password, and your full name, then click OK. Chameleon creates a mail directory for you. Click OK again.
4. Exit the Mail application, then double-click the Mail icon again. Log in with the username and password you just set up, then click OK. Your mailbox window appears, as shown in Figure 12-3.
5. Choose Settings | Network, then choose Mail Gateway from the submenu. The Mail Gateway dialog box appears.
6. Enter *smpt.interramp.com* for the Host, then click OK.
7. Choose Settings | Network, then choose Mail Server from the submenu.
8. Enter the host as *pop3.interramp.com*, enter your username as your account name, then enter your account password (not your e-mail program password). You don't need to enter anything in the Mail directory field.
9. Choose File | Save. You're now ready to send and receive e-mail.

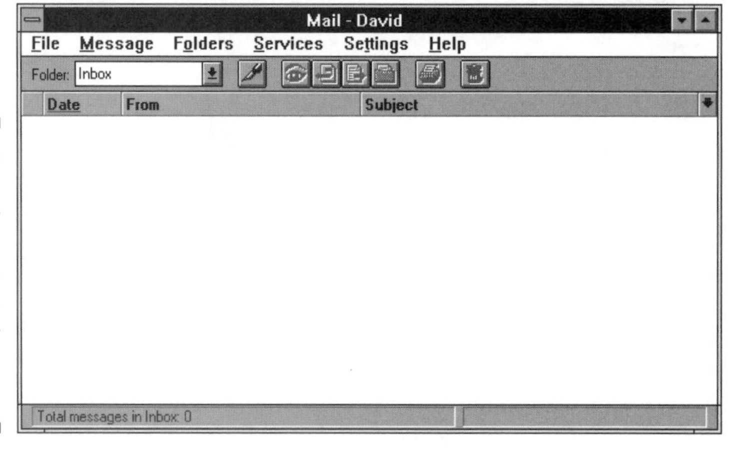

Figure 12-3: Once you create your mailbox, you can begin to configure it for managing your e-mail.

Setting Up NetManage's Network News Reader

Network News is the global bulletin board of the Internet, which includes thousands of newsgroups covering every imaginable topic. To set up the Chameleon newsreader program, do the following.

1. Click the NEWTNews icon. The same login dialog box that you used for e-mail appears.
2. Log in using the same name and password that you did for e-mail.

3. Click on Connect. A dialog box appears to enter the Network News host name, as shown in Figure 12-4.
4. Enter *usenet.interramp.com* and click OK. Internet Chameleon makes the connection to the network news server. After establishing a connection, NEWTNews fetches a list of all the news groups to which you can subscribe. This list takes a few minutes to download even using ISDN.
5. To subscribe to a newsgroup, scroll down the list or type the name of the newsgroup in the Group Name field. Click the plus (+) button to subscribe to the newsgroup. Selected newsgroups are indicated by the letter S to the left of their names. Continue adding as many newsgroups as you want. To deselect any newsgroup, highlight it and click the minus (-) button.
6. Choose File | Save.

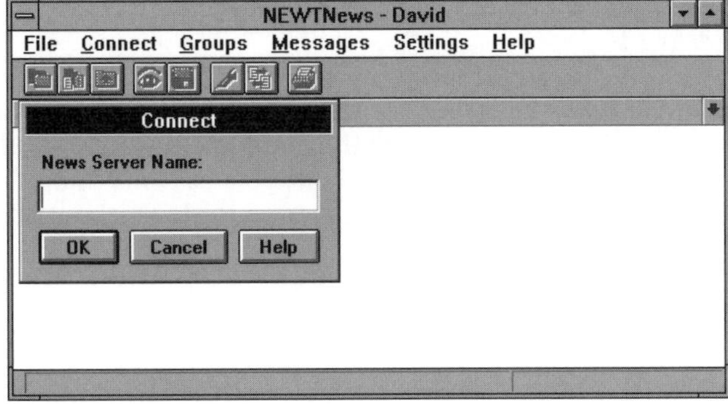

Figure 12-4: To get started with Network News, you enter the server name in the Connect dialog box.

Connecting to PSI

At this point you're ready to connect to the Internet via PSI InterRamp and ISDN. To connect to PSI, do the following.

1. Open the Internet Chameleon group.
2. Double-click the Custom icon. The Custom window appears, as shown in Figure 12-5.
3. Click the Connect command on the menu bar of the Custom window. The Log window appears. After the connection is complete, the Log window disappears. You're ready to surf the Internet!

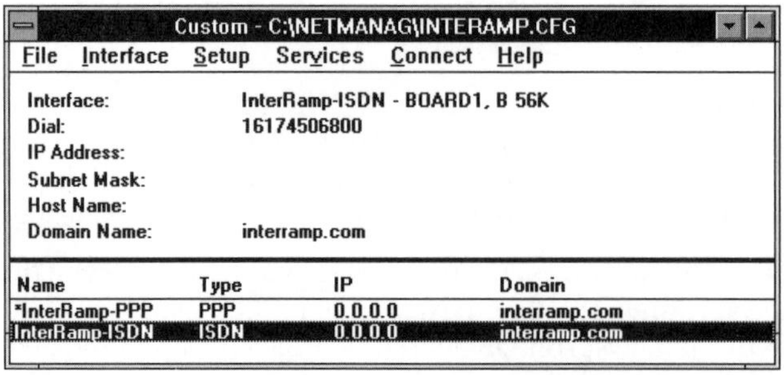

Figure 12-5: The Custom window appears with your settings in place, ready for you to connect to PSI.

Where to Now?

Beyond remote access, the last of the ISDN applications to consider is desktop video conferencing. This is one of the more exciting ISDN applications, which lets you conduct real-time video conferences from your computer. Next, you'll learn what desktop video conferencing is all about and you'll sample several leading packages.

Part IV

Face to Face via Desktop Video Conferencing

The 5th Wave **By Rich Tennant**

On March 3, 1992, Frank and Mona Tubman tried running multiple apps. through Windows on an OS/2 platform with their 286 DinQue PC.

The harddisk, seeking power from whatever source, began tapping appliances throughout the household electrical system, eventually sucking time itself from the wall clocks, thrusting the couple into an irreversible time-loop!

In This Part...

It's time to discover desktop video conferencing via ISDN. Desktop video conferencing systems use a video camera, a video capture board, an ISDN adapter, and software to turn your PC into a visual communication medium. In this part you learn the basics about how video conferencing works and how to use it effectively. You then experience desktop video conferencing via a walk-through of two leading systems.

Using these systems, you see first hand how you can share applications and work on a common whiteboard while also seeing the other person in a window on your screen. You learn about the standards that allow you to communicate across different systems and why desktop video conferencing systems that cost under $2500 are an appealing alternative to travel, how they extend your presence, and how they increase your productivity. You also learn how you can share applications and give presentations using programs like Microsoft PowerPoint.

Chapter 13

I Want My DVC (Desktop Video Conferencing)

In This Chapter

- What is desktop video conferencing?
- The components that make up a desktop video conferencing system
- What you need to do video conferencing
- An introduction to the leading DVC systems
- The elements of video conferencing style
- How to manage your video image

Of all the ISDN applications, desktop video conferencing is the one that most dramatically shows what ISDN can deliver. The convergence of affordable, standards-based desktop video conferencing systems combined with ISDN adds the powerful visual dimension to communications. For as little as $1000 and a PC, you can conduct a face-to-face meeting with someone down the hall or around the world. This chapter presents the big picture of the desktop video conferencing revolution and gives you a look at how it works.

The Downsizing of Video Conferencing

Video conferencing isn't new. Expensive, room-size systems incorporating specialized hardware and software components have been around for years. Large companies recognize them as productive tools and use them in everyday business. New video conferencing technology is downsizing these systems to inexpensive systems that work on a PC. Additionally, desktop video conferencing systems allow individuals to participate in meetings conducted in the room-size systems.

The enabling technology that allows desktop video conferencing to make a fast trek from the boardroom to the desktop include the advancement of data compression technology, the implementation of video conferencing standards, and ISDN. While developments in compression and standards helped bring desktop video conferencing to reality, it's ISDN that makes it affordable.

Seeing Is Believing

Video conferencing promises to bring the visual human element back into modern communications. It brings a valuable strategic tool to millions of individuals and small businesses for face-to-face meetings, team collaboration, brainstorming, training, and more regardless of the physical proximity of the participants. With the cost of desktop video conferencing falling under the $1000 mark, people from all walks of life can benefit from it. Desktop video conferencing:

- Extends your personal presence without the inconvenience and expense of travel.
- Speeds decision making and improves problem resolution with its show-and-tell capabilities.
- Allows specialists and experts to communicate with clients and team members regardless of their location.
- Reduces travel expenses for meetings. The cost for a coast-to-coast video call on a dialup network is less than $15 per hour.
- Makes meetings more productive. Each person at a video meeting has immediate access to both computer and paper documents, and can quickly involve other people who have special expertise.
- Enhances brainstorming and collaboration on the fly. Because video conferencing at the desktop is more intimate, it breaks down some of the formality barriers of a traditional in-person meeting.
- Creates flexibility in the work schedule by allowing people to work anywhere there are an ISDN connection and a PC.
- Allows telecommuters to attend meetings from home.
- Empowers virtual companies and organizations by enhancing management of geographically dispersed people.
- Provides new training and technical support options for customers.

Desktop Video Conferencing Systems

A complete desktop video conferencing system for the PC includes a video capture card, an ISDN adapter card, a small digital video camera, and video conferencing software. Figure 13-1 shows a typical desktop video conferencing system.

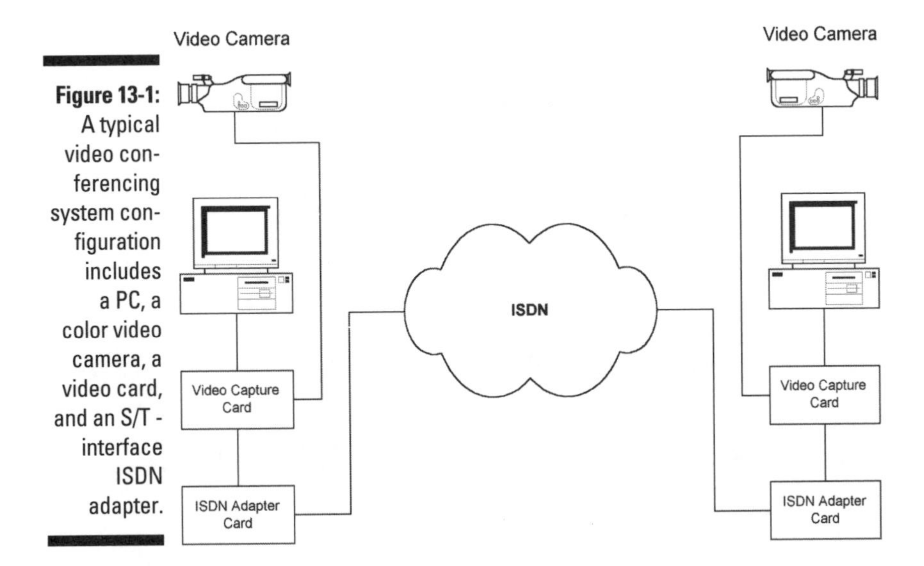

Figure 13-1: A typical video conferencing system configuration includes a PC, a color video camera, a video card, and an S/T-interface ISDN adapter.

There are two types of video conferencing tools: meeting tools and collaborative computing tools. Meeting tools allow you to present materials and perform file transfers. Collaborative computing tools allow users to share ideas and applications. From the desktop you can use video and audio conferencing tools along with sharing applications or using an electronic whiteboard — all happening simultaneously. Figure 13-2 shows how a typical video conference appears on a PC screen.

The leading PC-based video conferencing systems include Intel's ProShare Personal Conferencing Video System, Vivo Software's Vivo320, PictureTel's LIVE PCS 50, and AT&T's Vistium 1200. These systems are described in more detail later in this chapter. The following sections explain each of the key components of a typical desktop video conferencing system.

See Chapter 20 for more information about video conferencing vendors. ■

Figure 13-2: Desktop video conferencing uses your PC with ISDN and audio to create dynamic, on-the-spot conferencing.

Tools of the Video Conferencing Trade

Video conferencing systems typically include software for managing a video conference as well as for collaborative computing. Tools for managing a video conference include such features as adjustments for video and audio, address books for automating connections, and file transferring.

Whiteboard software lets participants illustrate points or develop ideas in the same way they would in a conference room with a whiteboard. Whiteboard features typically include annotation tools and tools to import information from other Microsoft Windows applications. Figure 13-3 shows the ProShare system's whiteboard application.

Application-sharing software lets participants work interactively by sharing the same application, such as Word for Windows or Excel. For example, you can share a spreadsheet so that both you and the other person can view and edit it.

Support Your Local Standards

For a video conferencing system from one vendor to work with another vendor's system, both systems must comply with the H.320 standard. This standard, established by the International Telephone Union (ITU), defines the interoperability between video and voice for video conferencing. Most desktop video conferencing systems comply with this standard, as do many room-size video conferencing systems. This means that desktop video conferencing systems can work with room-size video conferencing systems to conduct meetings.

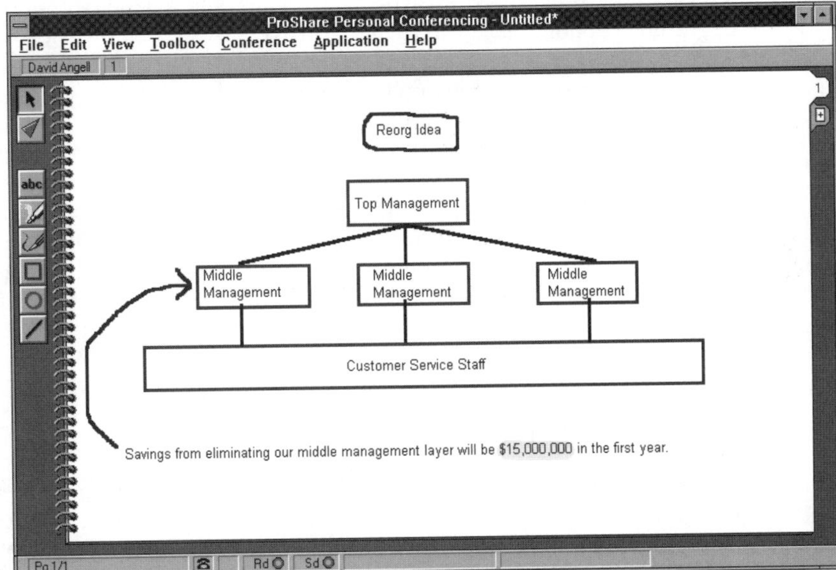

Figure 13-3: The ProShare system's whiteboard program provides a common ground for working with PC-based information during a video conference.

Beyond the visual element of desktop video conferencing, much of the interaction during a conference involves reviewing and marking up documents, and sharing information. A new standard called T.120 is emerging for exchanging and editing documents and images between participants in real time. When this standard is implemented in desktop video conferencing systems, you'll be able to use whiteboard and application-sharing programs across different products. Currently, you can't use the collaboration software from one DVC vendor with that of another. Look for most desktop video conferencing systems to support standards-based document processing.

Small Is Beautiful, Thanks to CODEC

At the heart of a video conferencing system is *CODEC*, which is short for *Coder/Decoder*. This is the engine that handles the compression and decompression of data for video conferencing. Data compression allows data-intensive video images to flow quickly through ISDN and your PC. PC-based video conferencing systems incorporate CODEC engine in three forms: a software-only solution, a hardware and software combination, and a purely hardware solution. Most of the current desktop video conferencing systems use the combination approach. However, the trend is toward software solutions that use the capabilities of faster Pentium and later CPUs, and high-speed graphics cards.

Software solutions depend on the CPU power of your PC to handle CODEC functions. For video conferencing systems using the software option, you need a Pentium system. The software-only package is less expensive and allows for upgrading capabilities via software improvements. The downside of the software CODEC is that it relies exclusively on your CPU. If you're using a CPU slower than a Pentium, the quality and speed of your video conferencing suffer.

The combination hardware and software CODEC solution shares the burden of compression between a processor on the video capture card and software. This middle-of-the-road approach takes pressure off a PC's CPU. However, it is less flexible in terms of future upgrades because the CODEC functions are in a processor on the video capture board. The purely hardware solution delivers the fastest capabilities for CODEC tasks, but it's the least flexible in terms of upgrading.

What's behind a Pretty Picture

The differences in the quality of desktop video conferencing systems largely boil down to their frames per seconds (fps) rate, resolution, and color bits per pixel. The higher the frame rate, resolution, and color bits per pixel, the better the quality of the video.

A *frame* is a single picture that, when pieced together with other pictures and displayed in a certain order, creates videos. *Frames per second* is the number of pictures flashed in a second to give the image the illusion of motion. Broadcast television presents a moving picture at 30 fps, which is based on the National Television Standards Committee (NTSC) standard. The NTSC standard defines how any television set in North America receives any transmitted broadcast television signal. Most desktop video conferencing systems deliver a video image at about 20 fps. At about 16 fps the illusion of motion changes to the perception of a series of frames.

Resolution refers to the number of pixels your monitor and card can display. Because of the huge data demands made by video, having a video image the size of your PC's full screen currently is not feasible. Image sizes for most desktop video conferencing systems fit within windows that range from 160x120 to 320x240 pixels in size. Pixels are picture elements, those little dots on screen that light up in different colors to make pictures. Standard video images measured in pixels are 160x120, 240x180, 320x240, and 640x480. The *bits per pixel* define the color depth of an image. The more bits supported, the better the color quality. Better graphics cards support 24-bit color.

Video Cameras (Say "Cheese")

Most video systems include a color video camera that attaches to your monitor or its own stand. The standard for digital cameras and video capture cards is the NTSC standard. Different video capture cards support different types of video cameras. For example, the Intel ProShare's video capture card supports both composite video via an RCA connector and S-VHS via a 4-pin mini-DIN connector. The advantage of an S-VHS port is that it allows you to connect a camcorder to your video conferencing system. In most cases, an S-VHS camcorder offers a better quality image as well as other sophisticated camera features.

If you have an S-VHS camcorder, get a video conferencing system with a video capture card that supports S-VHS. ■

The NTSC standard covers other countries besides the United States, Canada, and Mexico, including most Asian and South American countries. Much of Western Europe and other countries use the Phase Alteration Line (PAL) video format. Much of Eastern Europe, the former Soviet Union, and other countries use a video format called Séquential Couleur Avec Memoire (SECAM). ■

ISDN Adapter Cards

Most video conferencing systems include an ISDN adapter card with an S/T-interface. The ISDN adapter card allows your PC to communicate via ISDN. This card inserts into a 16-bit bus slot in your PC. The card connects via an eight-wire cable and an RJ-45 connector to an NT1 or NT1 Plus device. If you plan to use a remote access device, make sure you have another 16-bit slot available in your PC after installing your desktop video conferencing system.

What You Need for Video Conferencing

You need several key elements in place to work with video conferencing. They include a PC that meets certain requirements, a properly configured ISDN line, and an NT1 or NT1 Plus device. The following sections explain the specifics for each of these ISDN elements.

Your PC Configuration

Here's what you should have in terms of PC capabilities to get the most from video conferencing. In most cases you can get by with less, but you'll pay in performance.

- At least a 486DX2 CPU at 66MHz or faster; or better yet, a Pentium CPU at 60MHz or faster.
- An accelerated VESA-bus or PCI-bus graphics card that supports at least SVGA with 800x600 resolution and 256 colors.
- At least 8MB of RAM, but 16MB is better.
- Two empty 16-bit ISA bus slots, one for a full-length card and one for a three-quarter length card.
- Microsoft Windows 3.1 or 3.11, or Windows for Workgroups 3.11; and MS-DOS 5.0 or later.
- At least 6-15MB of available hard disk space.
- A 3.5-inch diskette drive to install the video conferencing software.

Graphics Cards and Monitors

Most video conferencing systems rely on your PC's graphics card and monitor. A graphics card sends pictures to your monitor. The better the quality of your graphics card and monitor is, the better is your video display. Here are optimal graphics card and monitor features for working with video conferencing.

- A 24-bit graphics card that can display 16.7 million colors, often called true color. This large number of colors is about the same number of colors that the human eye can differentiate.
- A dot pitch of .28 or smaller. The dot pitch is the distance between the pixel dots on the screen. The smaller the dot pitch is, the clearer is the picture.
- A noninterlaced monitor with 15-inch or larger screen (measured diagonally). A noninterlaced monitor's display has less flicker and looks better than that of an interlaced monitor.
- A VESA or PCI graphics card with 2MB of VRAM (Video RAM) that supports an extended graphic resolution of 800x600, 1024x768, or higher. VRAM is a special type of RAM that is faster than DRAM (Dynamic RAM).

ISDN Line Requirements

The ISDN line requirements for establishing a video conferencing system include the following.

- A single BRI line with two B channels and one D channel. The B channels can have data or mixed data/voice service. You can't have a voice-only B channel. You must use both B channels for video conferencing.
- An AT&T 5ESS Custom, a National ISDN-1 compliant, or a Northern Telecom DMS 100 telephone company switch. You also need to know the telephone numbers for your ISDN line and any SPIDs.

An NT1 or NT1 Plus Device

All the leading video conferencing products don't include built-in NT1, which allows you to add other devices to your ISDN connection. To use these system, you need an NT1 device, an NT1 Plus device, or an S/T-interface port on a remote access device. Typically, you'll use either an NT1 or NT1 Plus device that has an S/T-interface port for plugging in the RJ-45 cable from the video conferencing system's ISDN adapter card. The NT1 Plus device allows you to also plug analog devices into your ISDN line. See Part II for more information on NT1 and NT1 Plus devices.

Leading Desktop Video Conferencing Systems

The leading PC-based video conferencing systems within the $2500 and under class include Intel's ProShare Personal Conferencing Video System, Vivo Software's Vivo320, PictureTel's LIVE PCS 50, and AT&T's Vistium 1200. Most of these systems include a standard collection of desktop video conferencing features, but there are differences. Other desktop video conferencing products are available, but they're either more expensive or they don't ship as a complete system. The following sections provide a brief summary of the Big Four desktop video conferencing packages. The remaining chapters in this part take you on a test spin of Intel's ProShare, Vivo Software's Vivo320, and PictureTel's LIVE PCS 50 desktop video conferencing systems.

Chapter 20 provides a full list of desktop video conferencing vendors and information about how to save money buying a system. For example, you can buy an Intel ProShare system at the time you set up your ISDN service for only $999. ■

Intel stumbles out the DVC starting gate

The H.320 standard is now supported by all four of the vendors discussed in this chapter. The holdout for this standard was Intel. Intel tried unsuccessfully to create its own standard, called Personal Conferencing Specification 1.0, a specification that was incompatible with H.320. Intel now supports H.320 with a fix that, while not perfect, brings ProShare into the H.320 fold. Intel now ships two executable files, one for calling another ProShare system, and another for calling H.320-based systems. To make a video conference with others, you must know whether their system is a ProShare or an H.320-based system. Intel plans a single program for all systems in the future.

The Intel ProShare Personal Conferencing Video System 200

Intel's ProShare Personal Conferencing Video System 200 is the most affordable video conferencing system. It lists for $1995, but you can buy it at the time you get your ISDN service at a substantial discount. ProShare includes one of the best sets of collaboration software tools, including easy-to-use, built-in whiteboard and applications-sharing programs. The ProShare video conferencing system is the easiest of all the systems to install.

The biggest problem with the ProShare system was that it didn't support the H.320 standard (see FYI, above), so you could only use it with another ProShare system. But the ProShare system now supports the H.320 standard. Overall, the ProShare system is the best value for most PC users.

Vivo Software's Vivo320

The Vivo320 Personal Videoconferencing system is built around the IBM WaveRunner ISDN adapter card, a video capture board, anda camera. The interface design is built around the business meeting metaphor. You create and send video business cards, then conduct business presentations using standard business presentation software, such as Microsoft's PowerPoint. Vivo320 doesn't include a built-in whiteboard software program, but does include a third-party program called FarSite for Windows. Vivo320 complies with the H.320 standard and lists for $1995.

PictureTel LIVE PCS 50

PictureTel is no stranger to video conferencing. It's the market leader in room-size and portable video conferencing systems. Its new PictureTel LIVE PCS 50 desktop video conferencing system is designed to work as an extension of these systems for telecommuters. The PCS 50 includes good whiteboard and application-sharing tools. Unlike the other desktop video conferencing systems discussed here, the PCS 50 requires that you use one of PictureTel's own graphics card, which means you must replace your graphics card. The PictureTel LIVE PCS 50 lists for $2495.

AT&T Vistium Personal Video System 1200

The AT&T Vistium Personal Video System 1200 competes at the price level of the PictureTel LIVE PCS 50. It supports the H.320 standard and includes good whiteboard and application-sharing capabilities. On the downside, its BRI ISDN adapter card is not compatible with Windows for Workgroups if you're using its networking features. Another problem with the AT&T Vistium 1200 is finding a place to buy it. AT&T doesn't sell this system directly or through normal PC channels. You must go through their reseller channel, which is a maze for the lone customer trying to get the product.

Elements of Video Conferencing Style

Video conferencing is a visual medium that incorporates new ways to conduct one-on-one or group meetings. As a communications environment, it takes some elements from face-to-face meetings and some from telephone conversations.

Establishing a video conference is like dialing another PC using a modem. A video conferencing system makes a call via ISDN and establishes a connection. Each user typically uses a headset with a microphone and earphone. Most desktop video conferencing systems have a delay in the audio and video links. As you work with desktop video conferencing, you'll get into a slower mode of interaction. If you open an application on your PC, the video image slows down because the PC's CPU is using resources to open an application. Typically, you can minimize your local video window to reduce demand on your PC. A key point to remember in a video conference is not to get too far ahead of the system when speaking or moving around.

You can use desktop video conferencing to conduct one-on-one meetings or to participate in a multipoint video conference organized by a third-party service. To participate in a multipoint conference, you make a video conference call at a designated time to a designated video phone number. The conference organizing service manages the switching of participants' images. During multipoint conference, the audio signals from all video conferencing systems are combined so you can hear sound from all other systems. You'll see the video from the video conferencing system from which the user is speaking or the participant that the moderator has designated to speak.

AT&T Global Video Services provides a service called WorldWorx. This system lets you set up a multipoint video conference with other H.320-compliant video conferencing system users in the United States and most of the rest of the industrialized world. This system works like establishing a voice conference call. See Chapter 20 for more information. ■

Lights, Camera, Action!

Video conferencing thrusts you in front of a camera. As part of setting up a video conferencing system, you need to pay attention to lighting, camera placement, and even what color clothing you wear. The following sections present key concepts for presenting yourself in the best possible light in front of the camera.

Sculpting with Light

Three forms of lighting affect how you appear in your video image: key light, back light, and fill light. You use these to sculpt the subject of a video conference, which is you. The *key light* is the main light source that defines the basic shape of the subject. Typically, key light comes from above and in front of the subject, but somewhat to one side. *Back light* adds definition to the subject to counteract the key light. Typically, this is behind the subject. Fill light is placed behind the subject. *Fill light* is used to fill in any dark areas not covered by the key and back lighting. Figure 13-4 shows the placement of the key, back, and fill lights.

While you may not be able to control all the lighting elements for your video conferencing system, here are some basic lighting guidelines.

- Avoid strong back lighting. For example, avoid directing the camera toward a window or strong light source.
- Use fluorescent or white incandescent lights, if possible. Clear incandescent and older fluorescent lights tend to result in yellowish video images.

- Avoid direct lighting from the front, because it creates a flat look. The camera increases the contrast of your scene — dark areas gets darker, while light areas get lighter. This produces unnatural images.
- Make sure the lighting is adequate. Too little light results in grainy images.

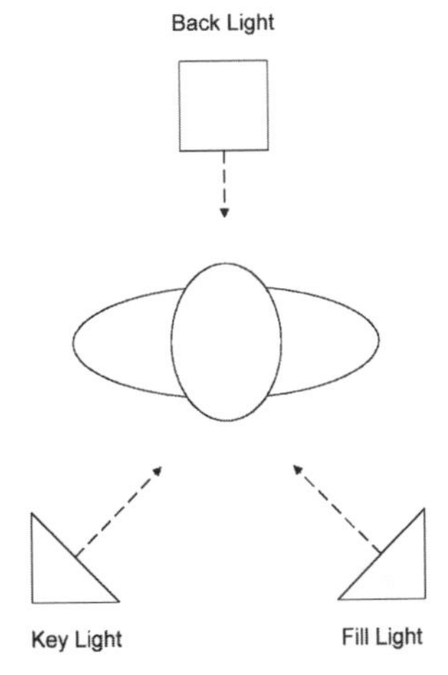

Figure 13-4: A well-chosen key, light, back, light, and fill, light layout improves your video image.

Backing Up Your Image

Beyond lighting, pay attention to your backdrop or background. Cameras catch whatever is behind you, so you may want to play around with backdrops to get the best possible environment. One of the best colors for a background is blue. To improve a background, experiment with making it darker and lighter. You can also have fun playing with different backdrops to create different views for remote viewers.

Clothes Make the Video Conference

One final area to pay attention to is the color of your clothing. Certain colors and patterns don't come across well on video. Here are some simple clothing guidelines for making the most of being in front of the camera.

- Dark or white clothing negatively impacts the lighting of your image.
- Reds bleed on screen.
- Patterns are difficult for the video conferencing system to constantly redraw.

Moving On

Now you have got a good idea of what desktop video conferencing is all about. The next chapters take you along in virtual video conferences using three of the four leading desktop video conferencing systems: Intel's ProShare, Vivo Software's Vivo320, and PictureTel's LIVE PCS 50.

"QUICK KIDS! YOUR MOTHER'S FLAMING SOMEONE ON THE INTERNET!"

Chapter 14

Video Conferencing with ProShare

In This Chapter

- How to install and get started with the ProShare video conferencing system
- How to conduct a video conference using ProShare
- How to use ProShare's computing collaboration tools

The Intel ProShare Personal Conferencing Video System 200 is the leading desktop video conferencing system. In this chapter you'll experience desktop video conferencing using Intel's ProShare system from installation to working with its collaboration tools within a Microsoft Windows environment.

About ProShare

Intel's ProShare Personal Conferencing Video System 200 is one of the most affordable video conferencing systems. Originally, ProShare did not comply with the H.320 standard, but bowing to industry and consumer pressure Intel is planning to make ProShare compliant with the H.320 standard. The ProShare system works on both a LAN and ISDN. The ProShare video system includes the following.

- A video capture card with both composite RCA and S-VHS ports.
- A digital camera that mounts on your monitor and connects to the video capture card using the composite RCA jack.
- A 16-bit audio/communication ISDN adapter card.
- A device with a microphone and an earphone that fits in your ear. This is one of the most annoying features of ProShare because it feels uncomfortable. You may want to purchase a better headset to use with ProShare.
- Collaboration software that lets you work with a whiteboard and share files in Windows applications.

Currently, the ProShare system lists for $1995, but you can get it for less as part of your order for ISDN service. You also can order ProShare directly from Intel, and they'll even establish your ISDN service for you. They maintain a center to help customers get up and running through the whole process. See Chapter 20 for more information.

Installing ProShare

Installing ProShare is surprising easy. Before you start, make sure you have a PC system that can support video conferencing, as explained in Chapter 13. You'll also need the following information about your ISDN connection to answer prompts during the setup of ProShare's software.

- Your ISDN line number or numbers.
- Any SPID numbers assigned to your ISDN line by the telephone company.
- The type of switch used at the CO.

Here's how to install ProShare:

1. Turn off your PC and remove the cover.
2. Choose two empty 16- or 32-bit slots and remove their cover plates. The two slots don't need to be next to each other.
3. Insert the ISDN adapter card into one of the slots. The ISDN board is the longer of the two boards that ship with ProShare.
4. Insert the video capture card into another slot. The video board is the shorter board.
5. Attach the camera cable to the back of the video card and camera, then plug the camera power cord into the back of the camera and power source.
6. Attach the headset to the HP and MIC jacks on the back of the ISDN board. Match the icons on the plugs with the board.
7. Mount the camera on the top of your monitor and attach the headset holder to the side of your monitor.
8. Attach the ISDN line to the Line jack on the ISDN board and connect the other end to an S/T-interface port in an NT1 or NT1 Plus device.
9. Turn on your PC and start Microsoft Windows.
10. Insert Disk 1, then from Program Manager choose File|Run. Type *A:\SETUP*, press Enter, and follow the prompts on the screen.

Getting Started with ProShare

When you start Windows, a ProShare Video Listening icon appears at the bottom of your Windows desktop. This is the ProShare program that sits in waiting for any incoming video conferencing calls.

To start ProShare, turn the camera on by sliding its door to the left. A green status light turns on. Double-click the ProShare Video icon at the bottom of the Windows desktop. The Handset window appears, and your image appears in the Local window. Figure 14-1 shows the ProShare Video window and identifies its main features. Table 14-1 identifies the key elements of the ProShare Video window.

Table 14-1: The elements of the ProShare Video window

Feature	*Description*
Call Status	Displays information about the status of your video call.
Handset	Includes the controls for operating the ProShare system.
Dial List	Shows a list of names or telephone numbers that you can automatically dial.
Dial	Executes the call of the number appearing in the Call Status box. Changes to Hang Up during a call.
Split	Separates the ProShare Video window into three separate windows: the Handset, the Remote, and the Local window. Changes to Combine after splitting the windows.
Options	Displays a Preferences dialog box for changing ProShare configurations.
Share Document	Executes programs for collaboration during a video conference, including whiteboard and application-sharing programs.
Remote	Displays the other person during a video conference call. Just below the Remote window is the Remote Tool Panel, which lets you adjust incoming video and audio.
Local	Shows your video image as it appears to the person at the remote location. When you start ProShare, your video image appears in this window. At the bottom of the window is the Local Tool Panel that lets you make adjustments to your video and audio.

ProShare uses ToolTip. This feature tells you about each button and feature when you put the mouse pointer over the on-screen object. Text appears in a yellow box.

Figure 14-1: The ProShare Video window is the main program window for ProShare. The left half is called the Handset. The area currently containing "Ready" is the Call Status box.

Turn off your screen saver to avoid interruptions during a connection. An active video in the ProShare Video window doesn't constitute activity for a screensaver.

Adjusting Your Video Image

The first thing you may need to do after starting ProShare is to adjust how you appear in the Local window, which is how you'll appear to the remote user in a video conference. You center yourself in the window by adjusting the camera. Clicking the Zoom button at the bottom of the Local window toggles between two camera views: close-up and wide.

Clicking the Camera Controls button at the bottom of the Local window displays the Camera Controls dialog box, as shown in Figure 14-2. These controls let you improve saturation, contrast, brightness, and tint. Table 14-2 explains the settings in this dialog box. To change settings, click and drag the sliders with your mouse. The changes take effect immediately as you move the slider.

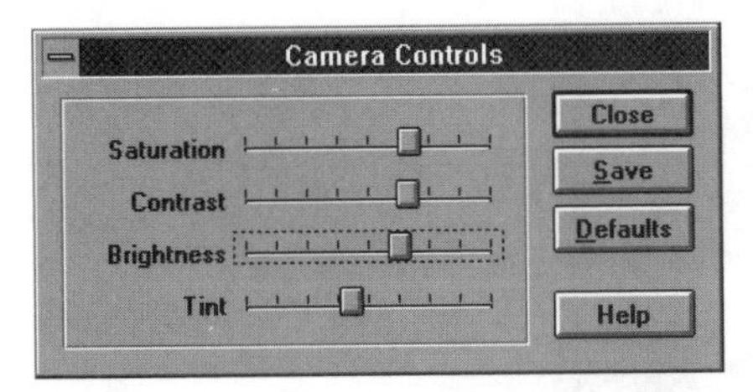

Figure 14-2: The Camera Controls dialog box lets you improve how you appear on video.

Table 14-2: ProShare Camera Controls dialog box settings

Setting	*Description*
Saturation	Changes the intensity of colors
Contrast	Adjusts the lightness or darkness of the image
Brightness	Adjusts the amount of light in the image
Tint	Affects the depth of colors from pale to deep

Managing Your Video Windows

You can split your video windows and arrange them so you have more room on your screen with the Split button on the Handset. The Handset, which looks like a telephone set, includes the controls for operating ProShare. When you click the Split button, the Remote, Local, and Handset windows become three separate windows, as shown in Figure 14-3. The Split button changes to the Combine button on the Handset.

Once the windows are separate, you can resize the Remote and Local windows, change any window to an icon, and move windows around your screen. Clicking the Combine button (previously the Split button) combines the Remote, Local, and Handset windows into a single window again. You can't change the combined window's size, though you can make it an icon. When you split the Local and Remote windows, an expanded Tool Panel appears under the Local and Remote windows. You can close the Tool Panel to save space on your screen by clicking the arrow button at the lower right side of the Local or Remote window.

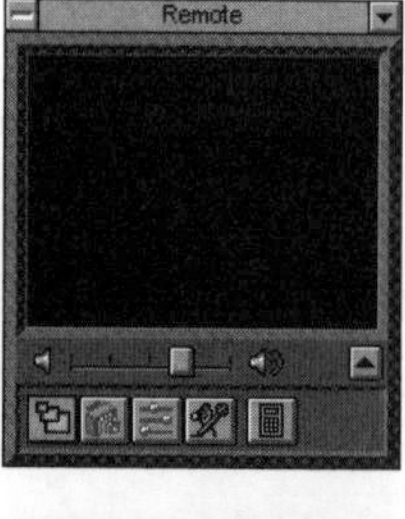

Figure 14-3: When you click the Split button on the Handset, the Remote, Local, and Handset windows become three separate windows.

Managing your windows becomes especially important when you're working with another application on your screen. For example, during a video conference in which you're working with a Windows application, you may want to display just the Remote window and minimize the Local and Handset windows, as shown in Figure 14-4.

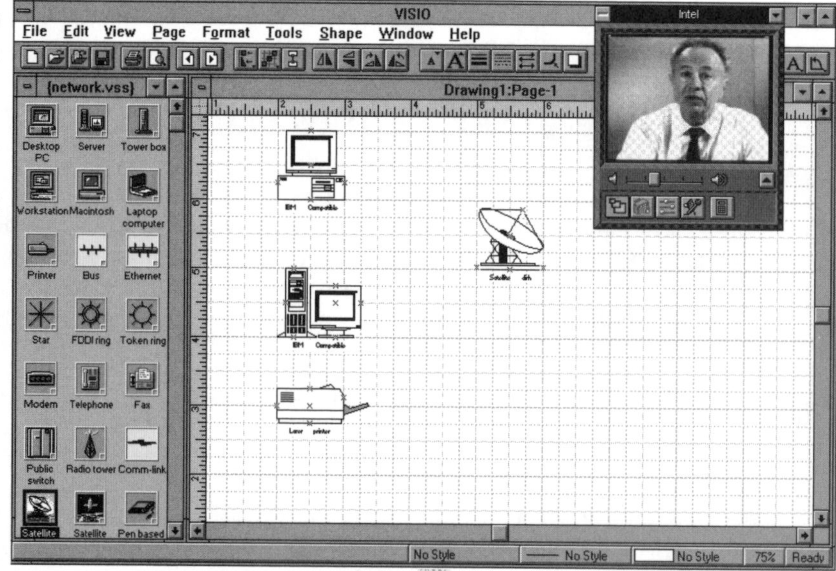

Figure 14-4: By splitting the windows, you can display just the Remote window, and iconize the Handset and Local windows to give you room for working on your PC screen.

Keep the Remote window under your camera during a video conference. This position allows you to look into the camera when talking with and viewing the other person so that person perceives you as looking at him or her. ■

Clicking the Resize icon at the bottom of the separate Local or Remote window displays a menu of choices for resizing the window, including Icon, Normal, and Large. The default size when all the windows are combined is Normal. The Normal window's dimensions are 160x120 pixels. The Large option expands the window size to 320x240 pixels. Figure 14-5 shows a comparison of the Normal and Large size windows.

Figure 14-5: The Local window using the Normal window size and the Remote window using the Large window size.

ProShare may automatically change the size of the video windows or minimize them to make the most efficient use of available resources. ■

Keeping the Remote Window on Top

You can specify that the Remote window always stays on top so you don't lose it during a video conference. This is handy when you're using ProShare's whiteboard or application-sharing features. To keep the Remote window on top for the current session, choose Always on Top from the Control menu (the box in the upper-left corner of the Remote window). To make the Remote window always appear on top for all video conferencing sessions, click the Options button on the Handset. The Options window appears. Choose the General option in the Categories list, then click on the Remote window checkbox.

Working with Audio

You can adjust the sound you hear in your earphone or speakers by moving the volume slider in the Remote window. The Speaker button (the button with a speaker and earphone on it) at the bottom of the Remote window lets you switch between your speakers and earphone.

If you want to use another headset with ProShare, set up ProShare, choose the Options button on the Handset. The Options dialog box appears. Choose Audio in the Categories list to display the Audio Options dialog box, as shown in Figure 14-6. The list of headsets supported in the Audio Options dialog box include several models from Plantronics, a leading vendor of telephone headsets. You can choose the Other setting that will work with other vendors' headsets.

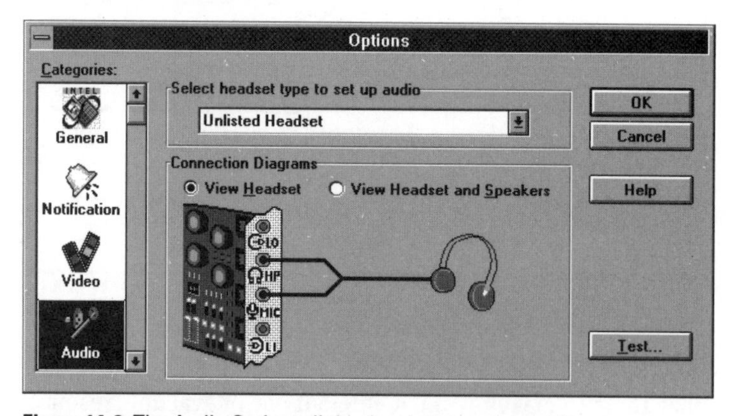

Figure 14-6: The Audio Options dialog box lets you use a driver for a different headset than the one that ships with ProShare.

Muting a Video Conference Call

ProShare lets you mute a video conference call. When you mute a call, you can still hear and see the other person. You're muting their ability to hear or see you, depending on the settings you choose. The other person sees a Mute notice when you mute the video. Click the Audio and Video Mute button (the button with the speaker and screen on it to the right of the Mute button) to choose one of the following settings: Mute Audio and Video, Mute Audio, or Mute Video. Click the Mute button to enable the Mute option you choose. When the Mute option is active, a red light appears on the Mute button. To turn off mute, click the Mute button again.

Making a Video Conference Call

You can make a video conference call by entering the phone number using the on-screen keypad or by entering the number and information in a Dial List so you can use it again without entering it. ProShare also offers a more sophisticated feature called an Address Book, which lets you include more information than the Dial List as well as preset other meeting parameters, such as linking files to share. Working with the Address Book feature is explained later in this chapter.

If you want to try a real number to make a video conference call, try 1-503-254-1369. This number connects you to a site at Intel that has a camera fixed on a fish tank. Another site you may want to try is the Intel Registration entry in the Dial List, which automatically registers you and plays a video clip of Intel CEO Andrew Grove. ■

Dialing a Video Conference Call

To make a down-and-dirty call without entering it in the Dial List, do the following.

1. Click the Headset button at the bottom of the Local window to display the Handset.
2. Move the pointer to the Call Status box and click.
3. Type your ISDN numbers. If there are two numbers for the ISDN line, enter the first number, a colon, then the second number. The colon acts as a separator. For example, 1-617-555-1212:1-617-555-1313. If you need to dial 9 for an outside call, make sure you enter it before the number, such as 9-1-617-555-1212.
4. Put on your headset.
5. Click the Dial button. After ProShare makes the connection, you see the person in the Remote window.
6. Click the Hang Up button in the Handset window after finishing your conference.

Speed Dialing via the Dial List

After you create a Dial List entry, you can make a video conference call simply by choosing the name of the entry. Here's how to set up a Dial List entry and make a video conference call with it.

1. Click the Edit button on the Handset (located under the Call Status box) to display the Dial List Setup dialog box, as shown in Figure 14-7.
2. Enter the name of the person you want to call in the First Name and Last Name boxes. Use the default Phone option in the Connection list for ISDN calls.
3. Enter the ISDN number in the Number box. If there are two numbers for the ISDN line, type the first number, a colon, then the second number. The colon acts as a separator. For example, 1-617-555-1212:1-617-555-1313. If you need to dial 9 for an outside call, enter it before the number, such as 9-1-617-555-1212.
4. Click the Add button to add the entry to your Dial List, then click Save to return to the Handset window.
5. Make sure your headset is on and click the Dial List button, which displays the list of entries in your Dial List.
6. Click the name of the person you want to call. The name appears in the Call Status box in the Handset window.
7. Click the Dial button. ProShare makes the connection and you see the other person in the Remote window.
8. Click the Hang Up button in the Handset window and ProShare disconnects the conference call.

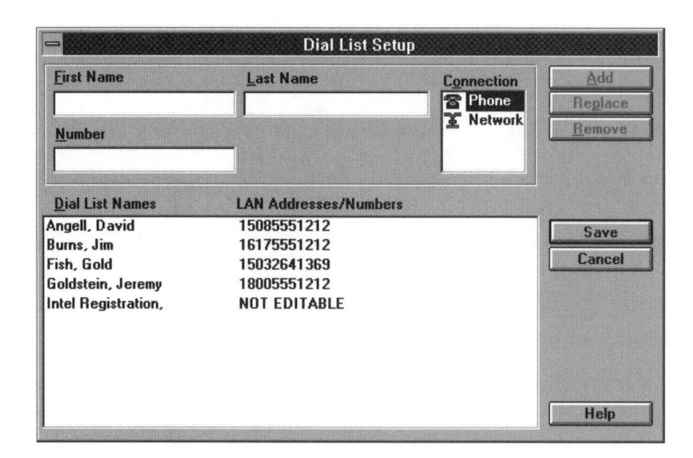

Figure 14-7: The ProShare Dial List Setup dialog box is where you enter information to automatic dial a number.

Answering a Conference Call

When you start Windows, ProShare automatically launches in listen mode. The listen mode runs in the background and notifies you when someone is trying to contact you with a video conference call. ProShare can notify you of an incoming call in two ways.

- You see an Incoming Call dialog box that tells you someone is trying to connect for a video conference. This is the default setting. Clicking Accept completes the connection and displays the person in the Remote window, ready for you to start your video conference. You can choose Decline, which displays a message to the caller that the call can't be completed.
- You can tell ProShare to automatically connect any incoming call. This lets any caller see you, even if you're not ready to take the call.

Shared Notebook Basics

ProShare's shared notebook tool is a multiple-page whiteboard on which you can write text and draw diagrams as well as bring in information from application files. When you're sharing the notebook, both you and the other person can view and make changes to the same information. Figure 14-8 shows the shared notebook window.

The shared notebook's toolbar includes a collection of tools for collaboration. You can take snapshots of different parts of the screen, import files, embed objects from other Windows applications, enter and highlight text, and draw pictures. The shared notebook also lets you transfer a copy of any file to the other person in the background while you continue to work in a notebook.

You can create different notebooks, which you can save and work on as an ongoing project. Each notebook has pages that you can leaf through as you would a traditional notebook. You can add, delete, copy, and print notebook pages. The shared notebook also lets you make private notes that the other person can't see.

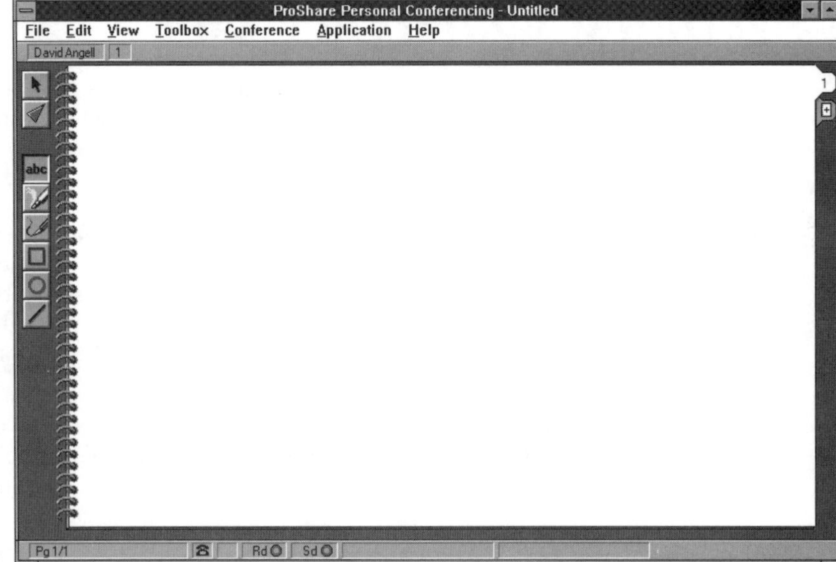

Figure 14-8: The shared notebook lets people conduct virtual meetings using a collection of useful tools.

Sharing Applications

During a conference, you can share any application on your or the other person's PC. When you share an application, both of you can view and edit a document together. Figure 1-4 in Chapter 1 shows sharing an Excel spreadsheet during a ProShare video conference. If you don't want the other person to edit the document, you can specify that he or she has view-only privileges. You must be connected to a video conference call to start sharing an application, and the person who starts application sharing must have the application installed on his or her PC. Here's how to work with application sharing.

1. Make the video conference connection.
2. Start the program you want to share.
3. Open the shared notebook by clicking the Share Document button in the Handset window. The shared notebook appears.
4. Choose Application | Share Application View from the share notebook menu. A dialog box appears listing the applications that are currently running on your system.
5. Click the name of the application you want to share, then click OK. The application window appears.
6. Work with the application as you normally would.

7. Save any changes you made as you normally would for the application. If the other person wants a copy of the file, you can transfer it by choosing File | Transfer. The File Transfer Status dialog box appears. Choose OK.
8. Choose Applications | Stop Sharing Application to end the application sharing session.

Taking Snapshots

You can take still-image snapshots of a video image in the Remote or Local window. You can add these pictures to shared notebooks, to the Address Book, or to any Windows application that supports bitmap images with the .BMP filename extension. Once you take the snapshot, you can copy it directly to the notebook, cut or copy it to the clipboard, or save it as a file. If your ProShare windows are combined, you can take only Normal-sized snapshots. If you split the windows, you can take snapshots in a larger size.

To take a snapshot and save it as a file, do the following.

1. Click the snapshot button in the Remote (or Local) window you want to capture. The Snapshot dialog box appears.
2. Choose the snapshot size from one of the following: Normal (160x120), Large (320x240), or Extra Large (640x480).
3. Click the Snap button to take a picture. The snapshot appears in the Snapshot window each time you click Snap. Figure 14-9 shows the Snapshot window.
4. Click on the Save to Disk button on the Snapshot window's toolbar. The Save Snapshot dialog box appears.
5. Choose the directory in which you want to store the file, enter a filename with a .BMP extension, then choose OK.

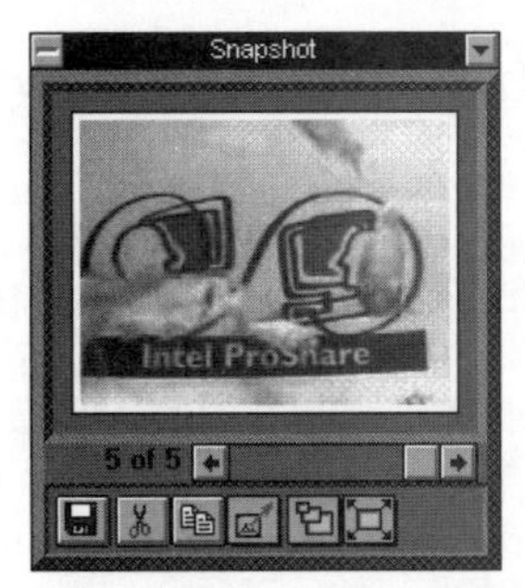

Figure 14-9: The Snapshot window shows the last snapshot you've taken and provides tools for working with the bitmap image.

Address Book Basics

ProShare's Address Book is a more powerful version of Dial List. It lets you add more detailed information about video conferencing participants as well as link files you want to share. When you use the Address Book to make a video conference call, the files you specify are immediately available for sharing.

Choosing Address Book from the Control Menu (the box in the upper-left corner) of the Handset window displays the Address Book window. Figure 14-10 shows the Address Book window and identifies its key features.

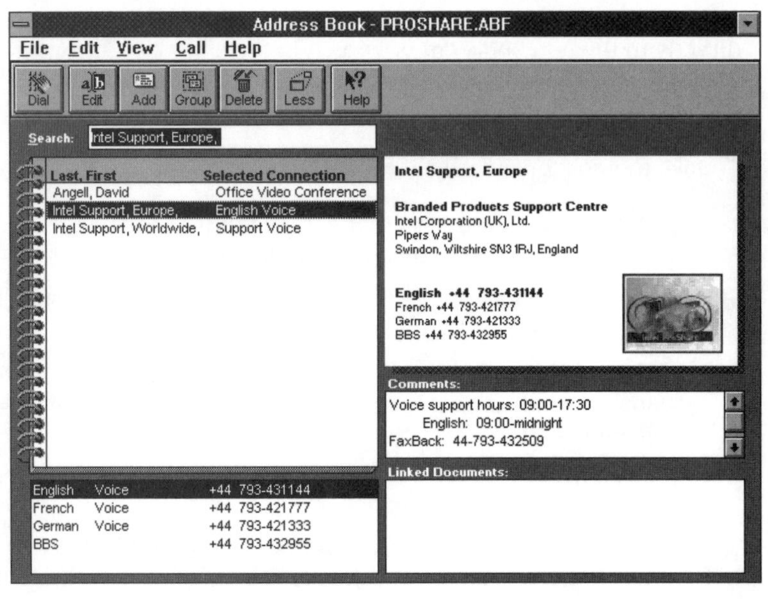

Figure 14-10: The Address Book window lets you manage an address book of video conference connections.

Adding an Entry to the Address Book

Adding a person to your Address Book involves entering optional address information, specifying the video number, then linking documents to that participant. To create an entry the Address Book, do the following.

1. Choose Address Book from the Control Menu in the Handset window. The Address Book window appears.
2. Click the Add button in the toolbar. A blank form appears in the Address Book window, as shown in Figure 14-11.

3. Type the person's first and last name. If you want to do so, fill in the other information. You can also choose the location of the number, such as home or office.
4. Click New to display the New Connection dialog box.
5. Choose the Phone or ISDN option in Select Transport and Video Conference in the Select Use list, then choose OK. The New Phone Connection dialog box appears.
6. Enter the Area Code, Local Number, Extension, Country Code, and Dialing Destination. The Dial As box at the bottom of the dialog box shows how ProShare will dial the number.
7. Click OK. You'll see the form again.
8. Choose OK to save the entry.

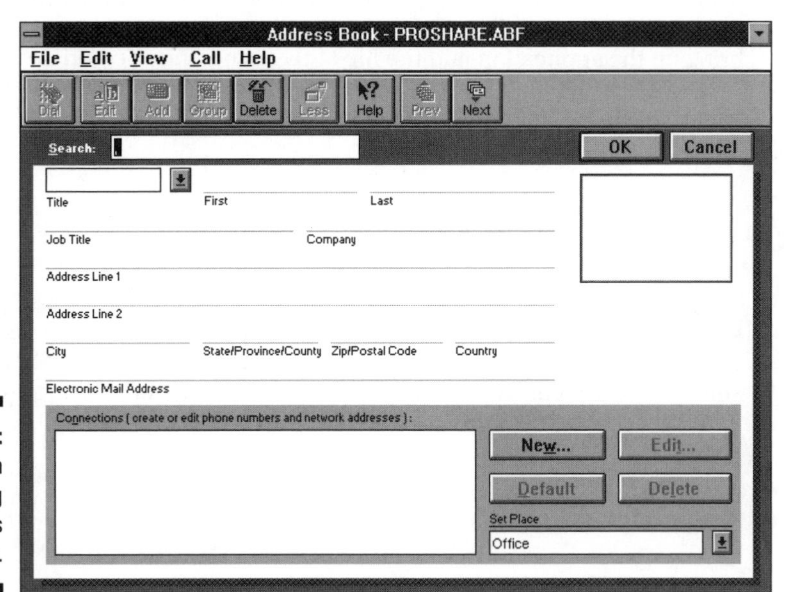

Figure 14-11: A blank form for adding an Address Book entry.

Adding Snapshots to Address Book Entries

You can add a snapshot to your Address Book entries by using ProShare's snapshot tool. The following steps explain how to add a snapshot to an existing Address Book entry.

1. Highlight the entry in your Address Book list and click the Edit button.
2. Double-click the rectangle on the right side of the entry form. The Browse dialog box appears.
3. Navigate to the directory containing the bitmap image file you want to add, highlight the file, and click OK. The picture appears in the entry.

Linking Documents to Address Book Entries

When you link a document, you connect another Windows application file to an Address Book entry. You can have up to 32 links per entry. This is a handy feature that saves time during a video conference call. After you make the connection, you simply double-click the filename in the Address Book to bring up the application and the file at the same time. To add document links to an Address Book entry, simply drag-and-drop files from the Windows File Manager into ProShare's Linked Documents list.

On to the Vivo320

You've experienced desktop video conferencing using Intel's ProShare Personal Conferencing System 200. Another leading desktop video conferencing system is the Vivo320, which is in the same price range as ProShare. It offer some interesting differences over the ProShare system. We turn to the Vivo320 in the next chapter.

Chapter 15

Experiencing Vivo320 Video Conferencing

In This Chapter

- How to install and get started with Vivo320
- How to conduct a video conference using Vivo320
- How to use Vivo320's Video Business Card and ScreenView tools

The Vivo320 Personal Videoconferencing system is, like the ProShare system described in the preceding chapter, another leading video conferencing system that runs in a Microsoft Windows environment. In this chapter you'll experience desktop video conferencing using Vivo Software's Vivo320 from installation to working with its collaboration tools.

About Vivo320

The Vivo320 Personal Videoconferencing system centers on the business meeting metaphor. It lets you conduct business presentations and exchange Video Business Cards. To support business presentations Vivo320 includes a handy tool called ScreenView, which lets you give live, continuously changing presentations. For example, you can run Microsoft PowerPoint on your system and give a slide show from Vivo320. After your presentation, you can send a file by dragging it to an icon.

Vivo320's Video Business Cards mimic standard business cards with pictures for making video conference call connections and exchanging with other video conferencing participants. Vivo320 doesn't include a built-in whiteboard software program, but includes a third-party program called FarSite for Windows.

Vivo320 complies with the H.320 standard. The product lists for $1995 and can be ordered directly from Vivo Software. Vivo will help establish your ISDN connection as part of your purchase. See Chapter 20 for more information.

The Vivo320 Personal Videoconferencing package includes the following components.

- A modified IBM WaveRunner ISDN adapter card that allows you to make calls and transmit sound, video images, and data over an ISDN line. It includes an ISDN cable that connects the ISDN card to an NT1 device.
- A video capture card with a composite RCA port and a proprietary camera port. Unfortunately, it doesn't include an S-VHS port for a camcorder.
- A digital video camera with two stands for positioning the camera either as a stand-alone unit or attached to the top of your monitor. The camera has a built-in microphone, so you can use your Vivo system as you would a speaker phone.
- Earphones and a speaker for listening to the other person's voice. When you want privacy, you can shut off the speaker and use the earphones.
- A setup and configuration program to install the Vivo320 software and configure the Vivo320.
- Vivo320 application software for working with desktop video conferencing in a Windows environment.

Installing Vivo320

The three things you need to run Vivo320 are a PC that meets certain requirements (see Chapter 13), an ISDN connection, and an NT1 or NT1 Plus device. Installing Vivo320 is a relatively painless process, but not as easy as installing Intel's ProShare. For your ISDN connection, you need the following.

- A single BRI line with two B channels and one D channel.
- Two data or voice/data B channels. You can't have a voice-only B channel.
- Information about the type of ISDN switch your telephone company uses for your BRI connection — an AT&T 5ESS Custom, National ISDN-1 compliant, or Northern Telecom DMS 100.
- Your ISDN telephone numbers.
- Any SPIDs supplied by the telephone company for your ISDN line.

The complexity of Vivo320's installation depends on what other cards you have in your system. The Vivo320 ISDN and video capture cards must use

DMAs, IRQs, and I/0 addresses not used by any other component in your PC. The first thing you should do before installing Vivo320 is take an inventory of all the DMAs, IRQs, and I/O addresses used by your current PC configuration. You can then compare them to the default settings for the Vivo320 cards, which you can then change if there are any conflicts. Vivo320 provides a handy installation worksheet to help you install it, and their technical support staff is excellent.

Here's how you install the Vivo320 system:

1. Compile all the information that you need for your ISDN line and enter it on the installation worksheet.
2. Start Windows, insert Vivo320 disk one in drive A:, choose File | Run in Program Manager, then enter *A:\SETUP*. Follow the dialog box prompts on your screen.
3. After completing the installation of Vivo320's software, turn off your PC.
4. Check the ISDN card hardware settings on the installation worksheet. If you need to make changes, follow the instructions on the worksheet to adjust switches on the Vivo320 ISDN card (the full-length card). You change the video capture card's I/O address using Vivo320 software.
5. Insert the ISDN and video capture cards in two available 16-bit ISA slots. They don't need to be next to each other.
6. Connect the camera to the back of the Vivo320 video capture card. Set up the camera using one of the two stands that ship with Vivo320. The monitor stand allows you to place the camera at the top of your monitor. The desktop stand allows you to place the camera next to your monitor. Regardless of which stand you use, make sure the camera lens is close to the area on the monitor where you will position the Remote Viewer window.
7. Connect the speaker to the back of the video capture card. Connect the speaker cable into the audio jack labeled OUT on the video capture card. Connect the earphones to the speaker cable. For the best audio quality, place the speaker as far as possible from the microphone. The microphone is built into the side of the camera opposite the focus knob.
8. Connect one end of the ISDN cable to the Vivo320 ISDN card in your PC and connect the other end to an S/T-interface port in an NT1 or NT1 Plus device.
9. Start Windows and double-click the Vivo320 icon in the Vivo Applications group. Vivo320 displays the Preferences dialog box for making adjustments to the camera. After making your changes and clicking OK, Vivo320 asks if you want to run the installation tests.

10. Click Yes to run the hardware and software tests. Vivo320 displays a message telling you the test will take up to a minute and that it will prompt you for input. Click OK. In a few seconds Vivo 320 asks if you can see yourself in both viewers and if you can hear your voice in the speaker.
11. Click Yes. If the tests ran successfully, Vivo320 displays a message to restart Vivo320.
12. Click OK. The setup program then asks if you want to make a test call to the Vivo Test Call System.
13. Click Now to verify that everything is running properly.

Getting Started with Vivo320

When you start Windows, the Vivo320 camera's power indicator lights. Make sure the camera lens cover is not in front of the lens. To start Vivo320, double-click the Vivo320 application icon in the Vivo Applications program group. The Vivo320 Control Bar appears, and the ISDN Port Manager icon appears at the bottom of your Windows desktop. Figure 15-1 shows the Vivo320 Control Bar.

Vivo320 uses ToolTip. This feature tells you about buttons and other features when you put the mouse pointer over an object. ■

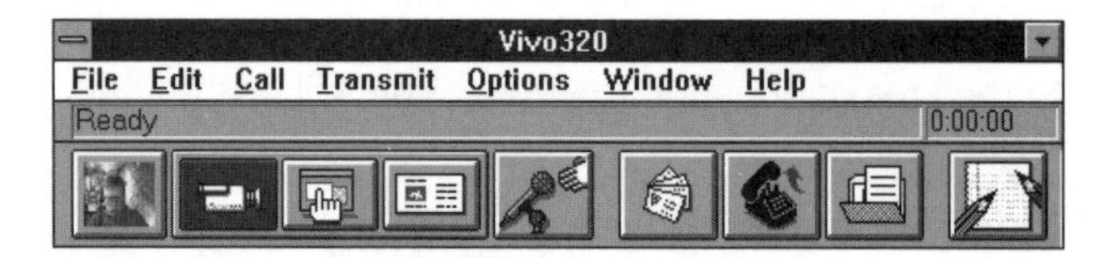

Figure 15-1: The Vivo320 Control Bar.

Turn off your screen saver to avoid interruptions during a video conference call. An active video in the Remote Viewer or Local Viewer window doesn't constitute activity for a screensaver. ■

Displaying and Adjusting Your Video Image

Clicking the Open Local Viewer button displays the Local Viewer window with you as the star attraction, as shown in Figure 15-2. You can make adjustments to how you look in the Local Viewer window by clicking the Camera Setting button, which is at the bottom right of the Local Viewer window. The Preferences dialog box appears. Click the Video tab to display the settings for making camera adjustments, as shown in Figure 15-3.

Figure 15-2: The Local Viewer window displays your video image.

Clicking Automatic Camera Exposure automatically configures the camera settings, including shutter speed and brightness. You can still adjust the lighting setting. The Automatic Camera Exposure control should give good results in most cases without the need for manual adjustments.

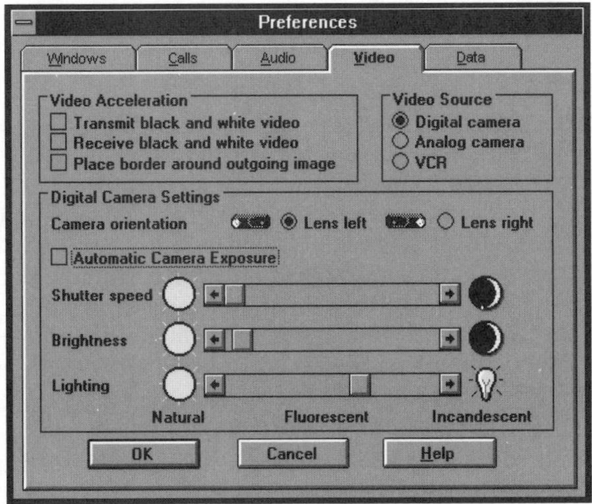

Figure 15-3: The Video tab in the Preferences dialog box lets you adjust the video camera and other Vivo320 settings.

You can improve video image quality for the Local Viewer and Remote Viewer windows by increasing the color resolution of your graphics adapter card. ■

Choosing the transmit black and white video and receive black and white video options in the Video Acceleration group of the Video tab displays your video images in black and white. Using black-and-white images reduces the amount of data transmitted, thus speeding up the video. You can also place a border around your image in the Local and Remote Viewer window by clicking the place border around outgoing image option in the Video Acceleration group. This option speeds up video transmission because it reduces the video image. Figure 15-4 shows the Local and Remote Viewer windows as they appear with the Border option active.

Figure 15-4: The Local Viewer window with a border. This option speeds up video transmission.

Managing Your Video Windows

Vivo320 lets you choose from two fixed window sizes for the Local Viewer and Remote Viewer windows, as shown in Figure 15-5. To expand the smaller window to the larger window, click the Maximize button (the up-arrow button at the top right side of the window). To shrink the window back to the smaller size, click the Restore button (the double-arrow button at the top right of the window). You can shrink a window to an icon by clicking the Minimize button (the down-arrow button after the window's title bar).

You can set other Vivo320 window configuration options by choosing the Camera Setting button or choosing Options | Preferences from the menus, which displays the Preferences dialog box. Clicking the Windows tab displays the available settings for working with Vivo320's windows. You can specify that the Local Viewer, Remote Viewer, and Control Bar windows always appear on top of any other windows on your screen. You can also change the default size of the Local Viewer and Remote Viewer windows to the larger size. Clicking save window positions on exit saves all Vivo320 window positions when you exit so they're available the next time you start Vivo320.

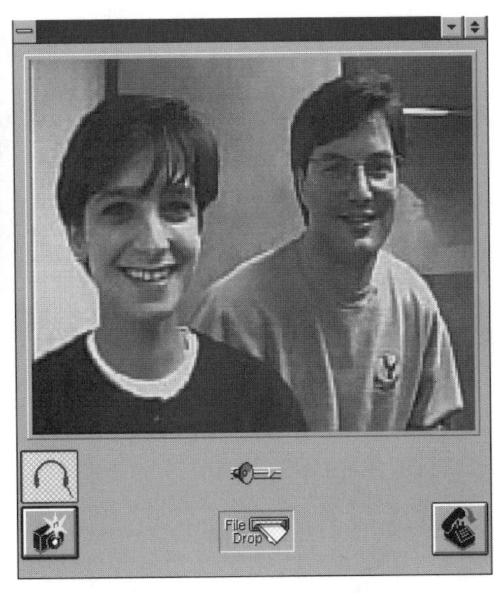

Figure 15-5: The two fixed window sizes for Vivo320's Local and Remote Viewer windows.

Working with Audio

Vivo320 includes several audio management features to work with the audio portion of your video conference. Here are Vivo320's tools for controlling the audio portion of your video conference.

- Clicking the Switch Microphone Off button on the Control Bar (fifth button from the left) turns off the microphone, which lets you talk privately during a video conference. Clicking the button again turns on the microphone.
- Clicking the Earphones button in the Remote Viewer window allows you to switch between receiving sound from the remote system through your speaker and your earphones.
- Dragging the Speaker Volume Adjustment in the Remote Viewer window adjusts the speaker or earphone volume.

Choosing the Audio tab in the Options|Preferences dialog box displays a collection of audio configuration options. The synchronize audio with video option in the Lip Sync group matches your audio with the video, which slows down audio. The minimize audio delay option lets the audio move faster but out of synch with the video. The Audio Output group lets you choose the audio default for the speaker or earphones.

Placing a Video Conference Call

Vivo320 lets you make video conference calls in two ways. You can place a call by dialing a number using the Dial Pad, or you can store a number as a Video Business Card for speed dialing.

Dialing a Video Call

Here's how to make a down-and-dirty video conference call using the Dial Pad.

1. Click the Open Dial Pad button in the Control Bar. The Dial Pad and the Business Card Tray appear, as shown in Figure 15-6.
2. Type the phone numbers in the Phone Number fields by clicking the Dial Pad numbers or by typing the numbers on your PC keyboard. Enter the number exactly as you would dial a voice number, including any prefixes, such as 1, the area code, then the telephone number. Remember, many ISDN lines require using two telephone numbers, one for each B channel.
3. Click the Dial button. Vivo320 dials the number, makes the connection, and displays the Remote Viewer window. The Business Card Tray and Dial Pad disappear.
4. After completing your video conference, click the Hang Up Call button in the Control Bar or in the Remote Viewer window.

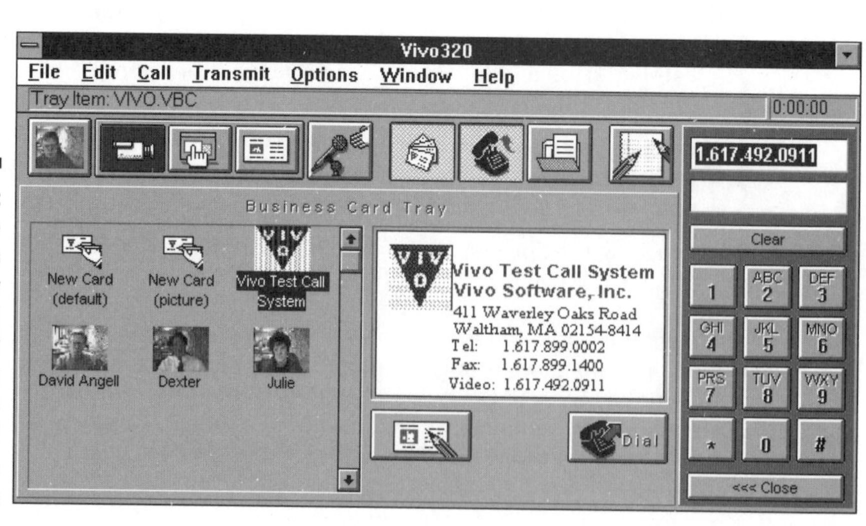

Figure 15-6: The Business Card Tray and Dial Pad appear when you click the Open Dial Pad button.

Making a Video Call via a Video Business Card

Making a video conference call using a Video Business Card is like direct dialing from a Rolodex card. You create Video Business Cards using the simple editor included with Vivo320, which is explained later. You can even add snapshots to your Video Business Cards. Here's an example of how to use a Video Business Card.

1. Click the Open Business Card Tray icon in the Vivo320 Control Bar. The Business Card Tray appears without the Dial Pad.
2. Double-click the Video Test Call System icon in the list on the left side of the Business Card Tray. A message box appears asking if you want to place the call.
3. Click Yes to make the call. Vivo320 makes the connection.
4. Click the Hang Up icon in the Remote Viewer window to terminate your call. The Remote Viewer window closes.

Answering a Call

Vivo320 must be running to notify you of incoming calls, either minimized or with the Control Bar visible. Whenever Vivo320 is running, the ISDN Port Manager icon appears at the bottom of your Windows desktop.

If you want Vivo320 to be ready every time you start Windows, add a copy of the Vivo320 icon to the Windows StartUp group. ■

Manual Call Answering

When a call arrives, the Incoming Call dialog box appears on your screen. To answer the call, click Answer. After a few seconds, the Remote Viewer window opens and you can begin speaking to the other person. Clicking Cancel displays a message to the caller saying there was no response from your system.

Unattended Answering

You can also set Vivo320 to use an unattended answer option. Choose Options | Preferences to display the Preferences dialog box, then click the

Calls tab, which displays the settings for incoming calls. Vivo320's Unattended Answer feature is handy for setting up an automatic information server for video conference callers. By selecting a screen region using Vivo320's ScreenView Tool, you can let remote users view a slide show anytime. When the unattended feature is active, Vivo320 displays the Incoming Call prompt; but if there's no response before the number of rings you specified, Vivo320 automatically answers the call.

If you use the Unattended Answer option without the ScreenView option, you may get caught on camera in a compromising situation—you just took a giant bite of a juicy submarine sandwich as a conference call connects! ■

Taking Snapshots along the Way

From Vivo320 you can take still snapshots of images in the Local Viewer and Remote Viewer windows. You can use these bitmap files, which end with the .BMP filename extension, for your own documents, presentations, and for Vivo320's Video Business Cards.

To take a snapshot of an image in the Local Viewer or Remote Viewer window, click the Snapshot button. The Document Tray appears with an icon of your image. You can continue to take snapshots, with each one showing up as an icon in the Document Tray, as shown in Figure 15-7. Double-clicking any of the image icons displays the image in the Local Viewer window.

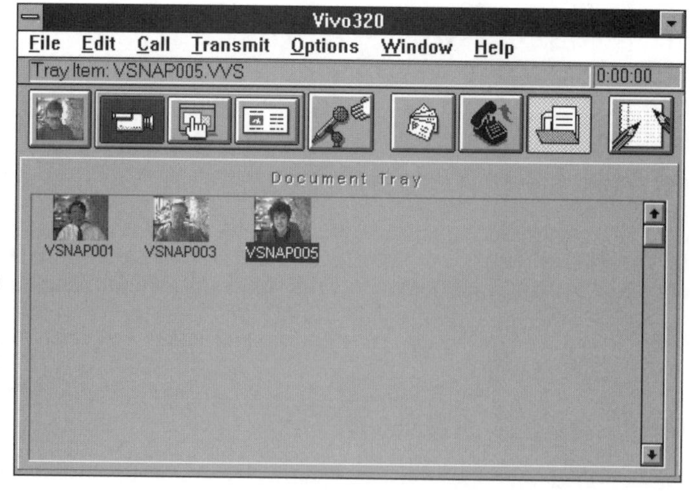

Figure 15-7: Document Tray snapshot icons.

Creating Video Business Cards

When you installed Vivo320, the setup program prompted you to enter information for your own Video Business Card. This Video Business Card acts like an electronic version of a standard business card, which you can exchange with other Vivo320 users. The Video Business Card tool also allows you to speed dial video conference calls using a business card stored in Vivo320's Business Card Tray.

You can choose from two default Video Business Card templates, one that includes a picture and one that doesn't. You can add additional call features to a Video Business Card, such as bringing up a specific application and file after you make a connection. Each Video Business Card appears in the Business Card Tray.

Here's how to create a Video Business Card that includes a picture.

1. Take a snapshot of the person in the Remote Viewer window for whom you want to create a Video Business Card.
2. Click the Open Business Card Tray button. The Business Card Tray appears.
3. Click the New Card (picture) icon. The template for the Video Business Card appears on the right side of the Business Card Tray.
4. Click the Name field in the Video Business Card template. The Business Card Editor appears, as shown in Figure 15-8.
5. Type the name and click OK.
6. Repeat the process for each field in the Video Business Card template.
7. Click the picture holder in the Business Video Card template. A dialog box appears for choosing the snapshot file you want to use.
8. Choose the file and click OK. The picture is added to your Video Business Card.
9. Click OK in the Business Card Editor to complete the card. The Video Business Card appears as a picture icon in the Business Card Tray.

You can add a picture of yourself for your business card by taking a snapshot of the Local window. Double-click your business card in the Business Card Tray. Click the picture holder in the template, choose the file, and click OK. Click OK to exit the Business Card Editor and save your business card.

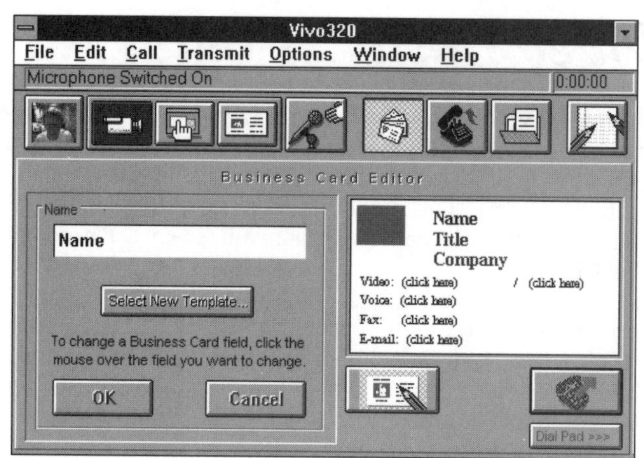

Figure 15-8: The Business Card Editor is used to enter information for each field in the card's template.

Sending Your Video Business Card Automatically

You can automatically send your Video Business Card to any video conference caller that connects to you, which places your business card in his or her Business Card Tray. Your business card is sent only once to each caller. This allows the caller to use your Video Business Card to speed dial you for future video conferences.

To send a Video Business Card automatically, choose Options | Preferences, then click the Calls tab. Click the Send copy of Video Business Card to remote system at call start option.

Making Presentations via ScreenView

The ScreenView presentation tool lets you show a portion of your Windows desktop to the person at the remote system. The ScreenView area is a live, continuously-updated image of an area of your desktop. You can show any area of your Windows desktop in which a graphic image, slide show, spreadsheet, document, or any other type of information appears. Figure 15-9 shows ScreenView in action during a video conference call. Keep in mind that ScreenView is not a two-way system. If both you and the person at the remote system want to edit the same document during a video conference, you must use a third-party document-sharing application.

Here is how to present an application window using Vivo320's ScreenView feature.

1. Start the application you want to use.
2. Start Vivo320 and open the Local Viewer by clicking the Open Local Viewer button.
3. Minimize or close any applications that are on your Windows desktop.
4. Click the Select and Transmit Screen Region button in the Vivo320 Control Bar.
5. Click the application in the Choose Application Window list.
6. Click OK. Vivo320 displays the ScreenView area in the Local Viewer window. The person at the remote system sees what you see in the Local Viewer window.

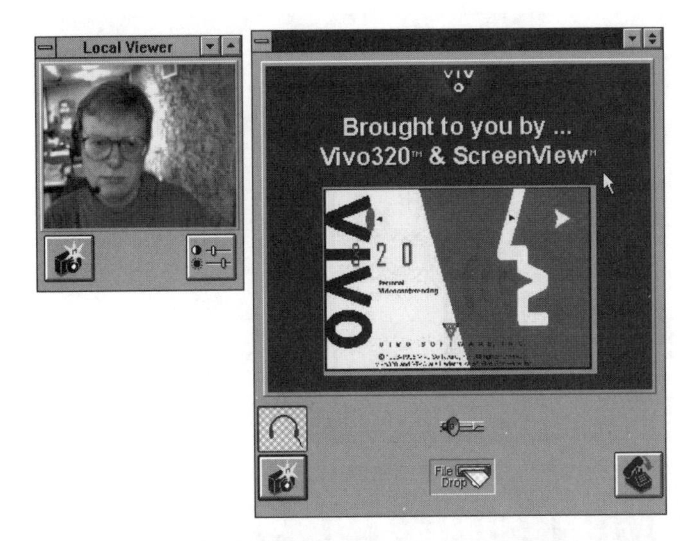

Figure 15-9: The Vivo320 ScreenView tool in action with a Microsoft PowerPoint slide.

Organizing Yourself with the Document Tray

The Document Tray lets you organize files you plan to use during a video conference in the same way you organize a presentation or documents to distribute during an in-person meeting. The Document Tray can hold files you plan to show or send, files sent to you, Video Business Cards, and snapshots you take within Vivo320. Figure 15-10 shows the Document Tray with assorted files in it.

To open the Document Tray, click the Open Document Tray button in the Vivo320 Control Bar (second button from the right). The Document Tray appears. You can add files to the Document Tray by dragging and dropping files from the Windows File Manager. Once you assemble your files in the Document Tray, make your connection.

Once connected, double-click an icon in the Document Tray to use Screen-View, or drag the file icon to the File Drop located at the bottom of the Remote Viewer window. File Drop automatically transfers the file to the other party.

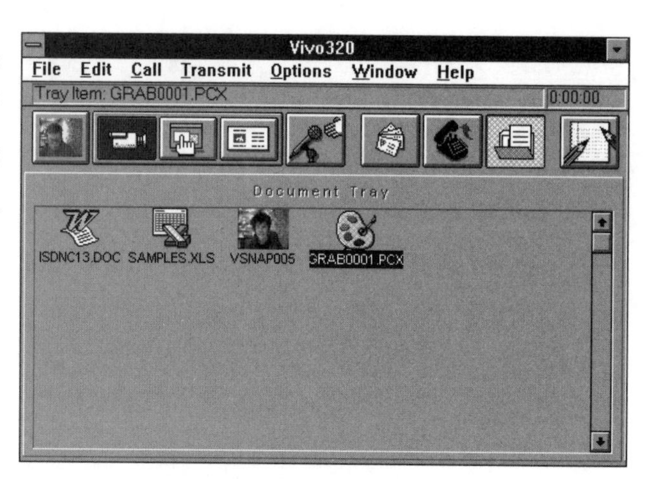

Figure 15-10: The Document Tray is a handy way to collect and organize your video conference call before you start.

On to the PictureTel LIVE PCS 50

The last desktop video conferencing system to check out is the PictureTel LIVE PCS 50. PictureTel is the leading vendor of room-size and portable video conferencing systems. Its entry into the desktop video conferencing market adds new capabilities for telecommuters whose companies already use larger PictureTel systems. The next chapter takes you for a test spin of the Picture-Tel LIVE PCS 50.

Chapter 16

DVC with PictureTel LIVE PCS 50

In This Chapter

- Installing and setting up the PictureTel LIVE PCS 50
- Making and receiving video conference calls with the PCS 50
- Working with the PCS 50's collaboration tools, including its chalkboard and application-sharing features
- Setting preferences for the PCS 50

PictureTel is a leader in the room-size and portable video conferencing industry. The LIVE PCS 50 is PictureTel's PC-based desktop video conferencing system. This chapter takes you on a video adventure using the PictureTel LIVE PCS 50.

About the PictureTel PCS 50

The PictureTel LIVE PCS 50 desktop video conferencing system is one of the more expensive desktop video conferencing systems, with a starting list price of $2495. The PCS 50 includes the standard equipment of other video conferencing systems, but it also includes its own VGA board that you must use in place of your own PC's graphics card. This is fine if you need to upgrade your video card; however, if you already have a newer video card, you can't use it.

PictureTel also sells a more powerful PCI-bus, VRAM video card from MGA. The PCS 50 has a full suite of collaboration tools, including chalkboard and application-sharing capabilities. It also includes remote control options that allow access to an unattended computer running the PCS 50. Lastly, it lets you place and receive audio-only calls.

PictureTel is a market share leader in the expensive, room-size video conferencing systems used by large organizations. But their experience in the PC market isn't so well established. For example, when I installed my PCS 50 I had problems getting the right IRQ and memory address settings. When I contacted their technical support, I got an engineer that was rude and terse. He offered no help in walking me through the solution. ■

Installing the PictureTel LIVE PCS 50

Installing the PictureTel LIVE PCS 50 involves two steps. The first step is installing the graphics adapter card and the PCS 50 card that includes the video capture, audio, and ISDN adapter functions. The second step is installing the PCS 50 software. The following sections explain how to install the PCS 50's hardware and software.

Installing PCS 50 Hardware

Installing the PCS 50 hardware includes setting up the graphics adapter card and the PCS 50 video board. You can purchase the PCS 50 with one of two graphics card options: an ISA bus card or a PCI bus card. The latter is the faster of the two cards. To install the graphics and PCS video cards, do the following.

1. Power off your computer and remove the graphics adapter card from your PC. Check the documentation for your system to see if you need to change any jumper settings on your PC's mainboard.
2. Insert the VESA-compatible VGA board into an available 16-bit ISA slot or insert the PCI bus VGA card into a PCI bus slot in your PC.
3. Power on your computer, start Microsoft Windows, and insert the disk labeled VAFC Drivers into the disk drive.
4. Choose File | Run in Program Manager and enter *A:\SETUP*. Follow the installation program's online instructions.
5. When the Board Selection dialog box appears, choose the PictureTel ISA Mapped option if you're using the VGA board for the ISA bus. Choose the 4-bit memory-mapped PCI board option if you're using the VGA board for a PCI bus.
6. Restart Windows and verify that your new adapter card is working.
7. Exit Windows and power off your PC.

8. Insert the PCS 50 card in a slot close enough to the graphics adapter card to use the VAFC cable. The card doesn't have to be in a slot right next to the VGA card.
9. Connect one end of the VAFC cable to the VESA advanced feature connector on the top of the PCS 50 board. Make sure the tab on the cable faces away from the socket plate end of the card.
10. Connect the other end of the VAFC cable to the VESA advanced feature connector on the VGA board. You can attach the PCS 50 board label to an empty socket plate to the left of your board if a space is available.
11. Plug one end of the video cable into the socket labeled Video on the camera and plug the other end into the socket with a camera icon on the PCS 50 board. Plug the power supply adapter into the back of the camera and plug the adapter into a power outlet.
12. Mount the camera on top of your monitor.
13. Plug the black connector with the speaker icon into the socket labeled with the speaker icon on the PCS 50 board. Plug the red connector with the microphone icon into the socket labeled with a microphone icon on the PCS 50 board.
14. Connect one end of the RJ-45 cable into the socket labeled ISDN on the PCS 50 board and plug the other end into an NT1 or NT1 Plus device. Power on your computer.

Installing PictureTel LIVE Software

To install the PictureTel LIVE software, you run the PictureTel LIVE Setup program. This program creates a directory named PCS50 for the PictureTel LIVE files. Before you install the PCS 50 software, make sure you have your ISDN line configuration information at hand. During the installation, the setup program prompts you for information about your BRI line, including the switch type and any SPIDs assigned to it by the telephone company. To install the PictureTel LIVE PCS 50 software, do the following.

1. Close any open Windows applications and insert the disk labeled Disk 1 Setup in your disk drive.
2. Choose File|Run in Program Manager and enter *A:\SETUP*.
3. Fill out the fields in the Registration dialog box and click Continue. The Type of Installation dialog box appears.
4. Click the Complete Installation button. The Installation Path dialog box appears.

5. Click Continue to use the default directory or enter a new path in the Install To field, then click Continue. The PictureTel LIVE ISDN Parameters dialog box appears.
6. Choose the switch type from the signaling dialog box and enter SPID numbers, if you were assigned any by the telephone company. Leave the default Not Used entries in the Number Type and Number Plan fields. Click the Continue button. The PictureTel LIVE PCS 50 Hardware Settings dialog box appears.
7. Choose the Auto Detect button to find the settings of your PCS 50 board, then choose Continue. The PictureTel LIVE PCS 50 Audio Device Selection dialog box appears.
8. Choose the Earpiece Only option, which is the unit shipped with the standard PCS 50 package, then click Continue. The PictureTel LIVE Node Name Selection dialog box appears.
9. Enter a unique name to identify your site to remote parties, then click Continue. The installation program copies files from disk 1 onto your hard disk, then prompts you for the next disk and so on until all five floppy disks are copied to your hard disk. After the installation program copies the files, it creates a PictureTel LIVE PCS 50 group window, as shown Figure 16-1.

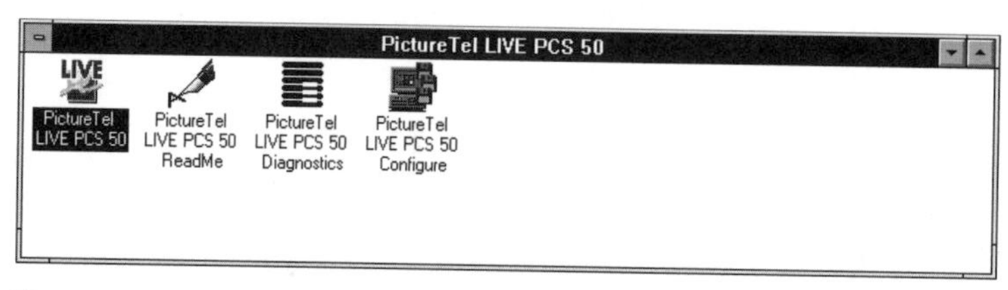

Figure 16-1: The PictureTel LIVE PCS 50 group window.

You can make changes to the software installation by opening the PictureTel LIVE PCS 50 Configure program icon. (Before you can open the Configure program, the PictureTel LIVE PCS 50 program must be closed.) When you open the Configure program, the same setup program you used to install and configure the PCS 50 appears. Follow the dialog boxes to find the one you want, make your changes, and exit the setup program.

Getting Started with the PCS 50

Before you start making and receiving calls, you may want to copy the PictureTel LIVE PCS 50 icon from the PCS 50 group to the Windows StartUp group. This loads the PCS 50 program each time you start Window so it can automatically answer video calls. To do this, simply press your Ctrl key while dragging the icon. By default the PCS 50 software starts minimized as an icon.

To start the PictureTel LIVE PCS 50 manually, open the PCS 50 group and double-click the PictureTel LIVE PCS 50 icon. The PictureTel LIVE window appears along with the Local Video window, as shown in Figure 16-2.

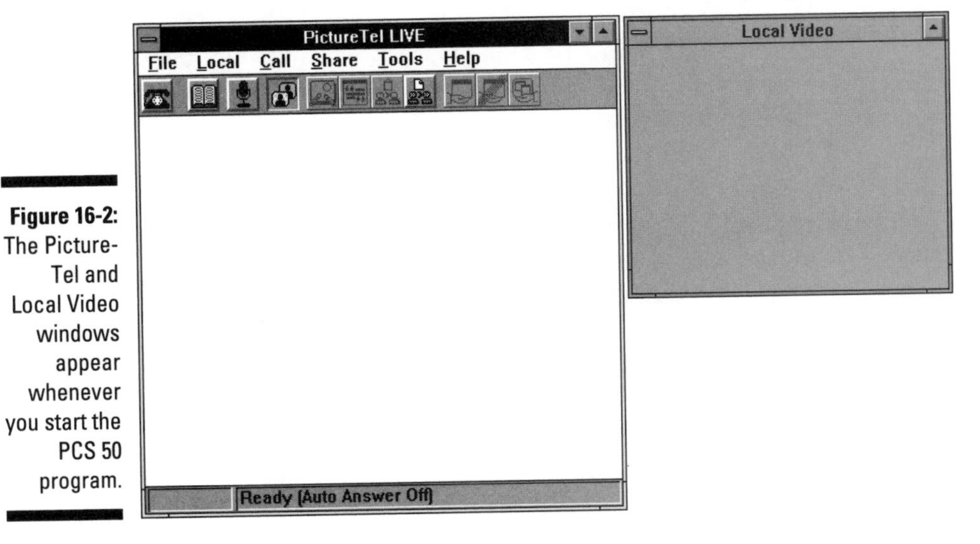

Figure 16-2: The PictureTel and Local Video windows appear whenever you start the PCS 50 program.

The PictureTel LIVE window includes the typical features of any Windows application, including menu and toolbars, work area, and status area. The work area of the PictureTel LIVE window displays the video from the remote site, while the Local Video window displays the local video image. Moving your mouse pointer over any button on the toolbar identifies the button in the Status bar. The dimmed buttons are for tasks that can only be activated during a video conference connection, such as the chalkboard or application sharing.

Unlike other desktop video conferencing systems, the Local Video window in the PCS 50 displays video from your camera at a full 30 frames per second. It looks great, but it's not what remote callers see. What they see is an image similar in quality to the image you see of them in your PictureTel LIVE window. ■

Placing and Answering Video Calls

Placing a video call is as easy as choosing a name from an on-screen phone book that also acts as an auto-dialer. Answering an incoming call is even simpler. The following sections explain how to place and answer calls.

Placing a Video Call

The quickest and easiest way to place a video call is by using the on-screen phone book. You can also call anyone not in your phone book with the dial pad. The entry you make in the dial pad can be quickly added to your phone book. The following steps explain how to enter a video telephone number in the dial pad, then add the entry to your phone book. This is how you build up your phone book.

1. Click the call button (the telephone icon) on the PCS 50's toolbar, or choose Call | Dial Pad. The Dial Pad dialog box appears, as shown in Figure 16-3.
2. Enter the first phone number in the First Phone Number field. Don't press Enter, because that executes the call.
3. Click the Second Phone Number field or Press Tab. The phone number you entered in the First Phone Number field appears. You must enter two numbers, even if they are the same. You need one entry for each B channel. If the second phone number is different, enter the new number.
4. Choose the Channel Rate. In most cases, a long distance, circuit-switched data call uses 56 Kbps.
5. Click the Add to Phone Book button to save the numbers to the Phone Book. The Add Entry to Phone Book dialog box appears, as shown in Figure 16-4.
6. Type the first name, last name, and company name. You must fill out at least the first name field.
7. Click OK. The first time you create an entry for the phone book, you're prompted to name a phone book file. Enter a filename for your phone book and choose OK. The Dial Pad dialog box appears.
8. Click the Video Call button to make your call. After the call is established, the person at the remote site appears in your PictureTel LIVE window.
9. To end a video call, hang up by clicking the call button or choosing Call | Hang Up. If the other party hangs up, your system automatically hangs up too.

To place a video call using the phone book is easy. Click the phone book button in the toolbar or choose Call | Phone Book. The Phone Book window opens, as shown in Figure 16-5. Double-click a name to make the call. The connection is made, the call button in the toolbar shows the headset lifted off the hook, and the PictureTel LIVE main window shows the video from the remote site.

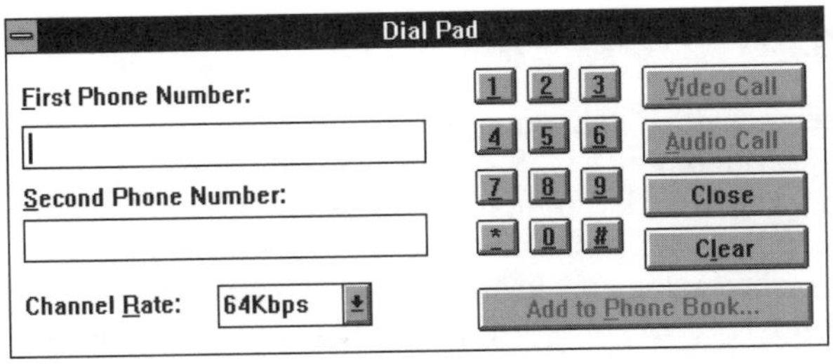

Figure 16-3: The Dial Pad dialog box lets you enter a phone number and make the call, as well as add the entry to your phone book.

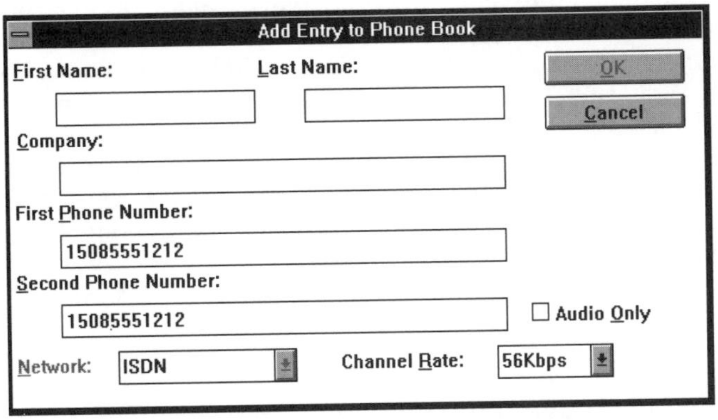

Figure 16-4. The Add Entry to Phone Book dialog box is used to add entries to your phone book.

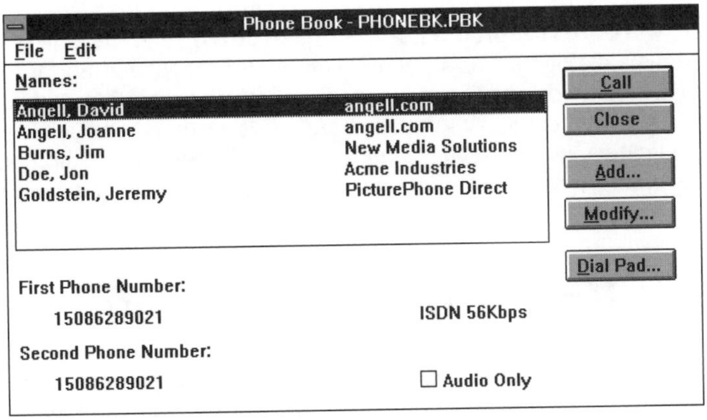

Figure 16-5: The Phone Book dialog box includes all the entries in your phone book.

Answering a Video Call

When you receive a video call, you typically answer it manually. When the PCS 50 detects an incoming call, it displays an Incoming Call dialog box. To answer the video call, click the Answer button. If you don't want to answer the call, click the Reject button. You can also set up the PCS 50 to automatically answer any incoming call. In most cases, you won't use this feature unless you want to set up your PC for remote control. Setting up the PCS 50 for remote control operation is explained later in this chapter.

Making and Receiving Audio-Only Calls

You can use the PictureTel LIVE PCS 50 to make audio-only calls using the same tools used to make a video conference call. The only difference is that when you create an audio-only entry in the Add Entry to Phone Book, you click the Audio Only checkbox. If you make a voice-only call from the Dial Pad, click its Audio Call button.

You answer an audio-only telephone call in exactly the same way that you do for a video call. When the PCS 50 detects an incoming audio call, it displays a dialog box indicating the call is an audio call. Click the Answer button to receive the call or the Reject button to decline it.

Setting Preferences

The PCS 50 Preferences dialog box lets you customize how the PictureTel LIVE system works. To set preferences, choose Tools | Preferences to display the Preferences dialog box, as shown in Figure 16-6. At the left under the Category list are five categories of preferences: General, Video, Audio, Network, and LIVE Share. Click on a category to change any of its settings. The settings for the selected category appear in the middle of the Preferences dialog box. After making changes to one or more categories, choose OK to save your changes.

Adjusting Your Video and Audio

You can make a variety of adjustments to the video and audio you send and receive. Here are some key controls for adjusting video and audio.

- Enlarge the Local Video or PictureTel LIVE window to full size by clicking its Maximize button.
- Mute your audio so the other party doesn't hear you by clicking the Microphone button on the toolbar. When your audio is muted, the Microphone button changes to the Mute Microphone button. To resume audio, click this Mute Microphone button.
- Adjust the audio from the remote party with the volume control at the left side of the PictureTel LIVE window. Simply drag the pointer up or down.
- Adjust the resolution of your video. Changing to a lower resolution reduces the data demands on your ISDN line so you can get better results working with collaborative tools or file transfers. To change the resolution setting, choose Options | Preferences, click the Video icon in the Category list, then click Quarter CIF in the Video Resolution settings group.
- Choose between smoother motion or a sharper picture. If your video conference includes very little motion, you can choose the Sharper Picture option in the Video category of the Preferences dialog box.

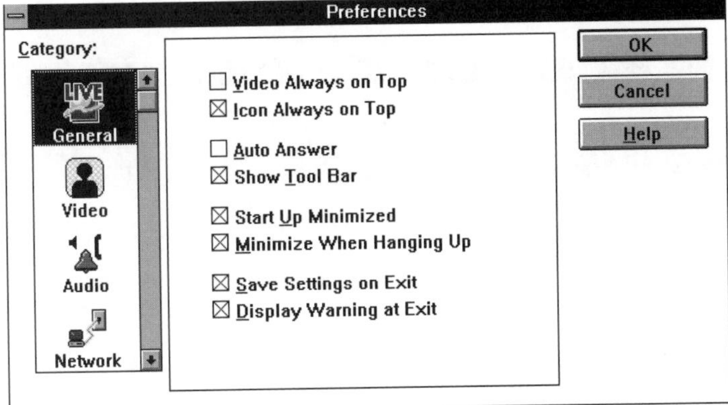

Figure 16-6: The Preferences dialog box lets you change default settings for the PCS 50.

Working with the Chalkboard

The chalkboard is a collaboration tool that lets you interact with the remote party using a visual workspace on both of your computer screens. You can only use the chalkboard when you're connected to another PictureTel LIVE system, and it automatically closes when a video call ends.

To use the chalkboard, click the Chalkboard button on the toolbar or choose Tools | Chalkboard during a video conference call. The Chalkboard window appears, as shown in Figure 16-7. The toolbar gives you access to all available tools. The open area of the window is the collaboration area, where you can perform any of the following tasks.

- Open and display a graphics file with a .BMP, .PCX, or .TIF extension.
- Paste a graphic or text from the Windows Clipboard.
- Mirror another application window, which appears in the background so you can work in that application as well as use the tools of the chalkboard.
- Call attention to objects on the chalkboard by identifying them with a pointer.
- Draw lines of different widths and forms, including freehand, boxes, and circles.
- Type text in different colors.
- Use an eraser tool to remove text or a drawing from the chalkboard.
- Print a chalkboard or save its contents.

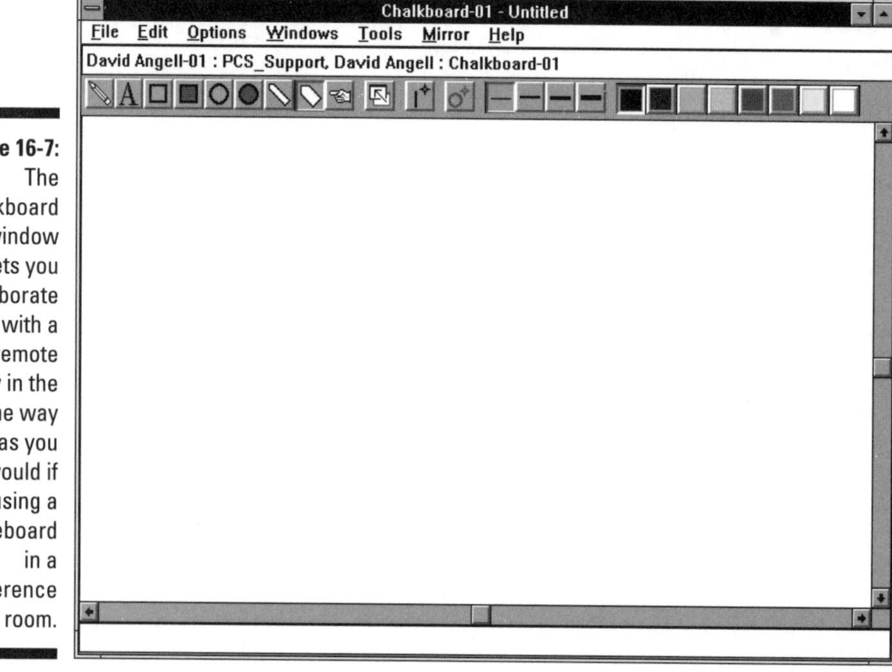

Figure 16-7: The Chalkboard window lets you collaborate with a remote party in the same way as you would if using a whiteboard in a conference room.

Sharing Applications

The LIVE Share program lets you share any Windows application during a video conference call. Both you and the other party can work with the same application regardless of whose system it's on. When you share an application, an exact copy of the application's window appears on the other party's Windows desktop. One party controls the application while the other party watches. As long as any application remains shared, you and the other party can alternate between controlling and viewing. While you're viewing, you have no control over the shared application or over anything on your desktop. To work with a shared application, you must first take control. The following steps explain how to share an application during a video conference call. You can share several applications by applying these steps to each one.

1. Open the Windows application on your desktop.
2. Click the Share button on the toolbar or choose Application | Share Application. The cursor changes to a share shape when you move it over an application window.
3. Click to share the application. A copy of the application window appears on the other party's screen. A tag with your system's name appears at the right corner of the window to identify that the application is on your PC.
4. Click your mouse button or press a key on the keyboard to take control of the application. Your status line changes to In Control. You remain in control until the other party clicks his or her mouse. When the other party takes control, your status line changes to Viewing.
5. After you finish sharing an application on your system, take control by clicking your mouse button, then click the Unshare button on the toolbar.
6. The cursor changes to an unshare shape when you move it over a shared window. Click the window you want to "unshare."

Transferring Files

The PCS 50's file transfer features let you send and receive files during a video conference. Files are packages that you send to the remote party using a simple file transfer program. The following section explain how to send and receive files.

Sending File Packages

To send files you create a package that includes a one-line subject line, one or more files, and an optional message of any length. You then send this package to the remote site. You can't create a file package unless you're connected. The following steps explain how to send files.

1. Click the File Transfer button on the toolbar or choose Tools | File Transfer. The File Transfer window appears, as shown in Figure 16-8.
2. Click the Create button. The File Transfer — Send dialog box appears, as shown in Figure 16-9.
3. Type a one-line subject description in the Title field.
4. Click the Attach button. The standard Windows Open dialog box appears. Navigate to the drive and directory, then select one or more files to attach. To select a single file, click its name; to select multiple files in the same directory, press Ctrl and simultaneously click their names.
5. If you're sending a large file or a bunch of files, click the Compress files checkbox in the File Transfer — Send dialog box. The remote system automatically expands compressed files.
6. Optionally, you can type a message in the Text field to accompany the files.
7. Click the Send button. The File Transfer — Send dialog box closes and a progress message appears in the File Transfer window showing the percentage of the files transferred.

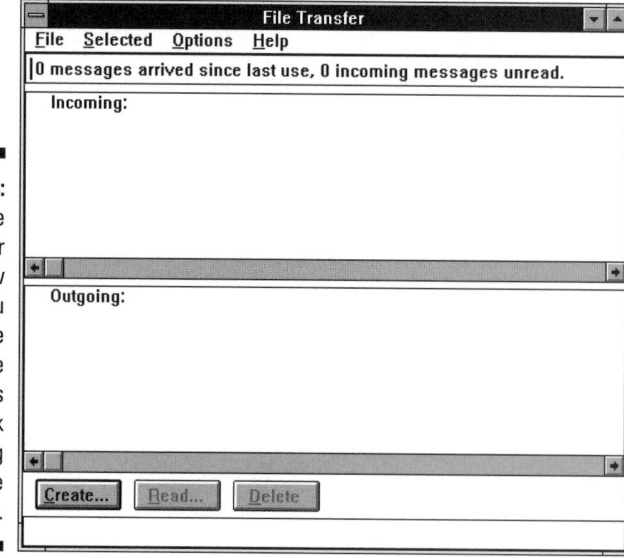

Figure 16-8: The File Transfer window lets you assemble your file packages and unpack incoming file packages.

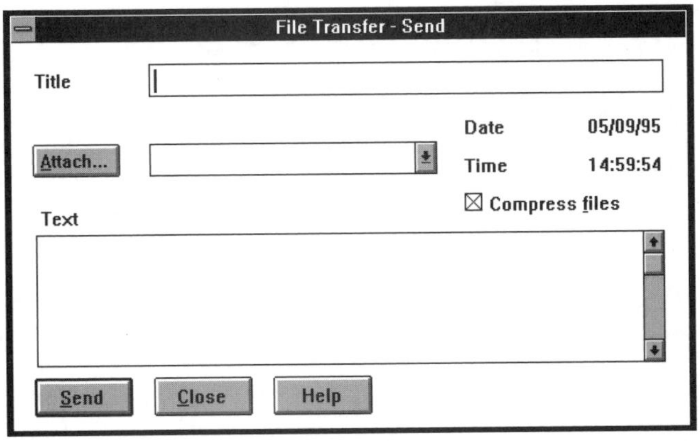

Figure 16-9: The File Transfer — Send dialog box is where you create file transfer packages.

Receiving File Packages

When you receive file packages from a remote site, your system lists them in the Incoming section of the File Transfer window. The PCS 50 stores the file packages you receive in a temporary default directory, which you can move to any directory on your system. You can also define the maximum size of incoming file packages that you will accept. To set the storage location and space options in the File Transfer window, do the following.

1. Choose Options | File Transfer Store in the File Transfer window. The File Transfer Store dialog box appears, as shown in Figure 16-10.
2. Type a new path to change the directory location for storing incoming file packages.
3. To change the amount of disk space reserved for incoming file packages, enter a new size. The default is 2000K (i.e., 2MB).
4. Click OK.

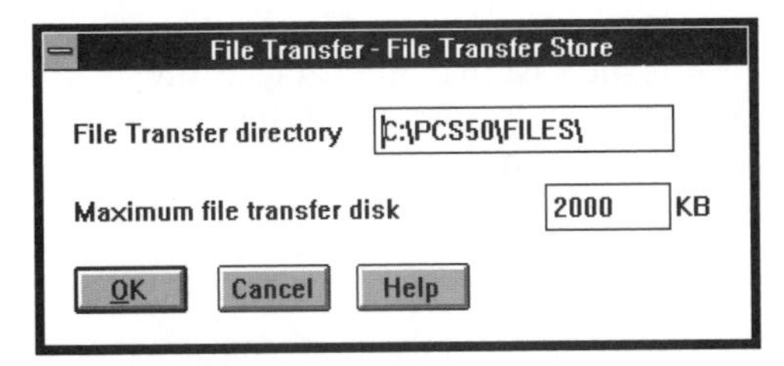

Figure 16-10: The File Transfer Store dialog box lets you change the default directory in which to receive incoming files as well as the maximum incoming file size you will accept.

Remote Control via the PCS 50

You can set up your system for unattended remote access. This allows others to access your system and share its applications — including Program Manager. Because they can control Program Manager, remote users have complete control over your system. To protect your system, the PCS 50 requires remote users to enter a password.

This remote control is a nice solution for telecommuting to your office system. The following steps explain how to set up the PCS 50 for remote control access.

1. Choose Tools | Preferences to display the Preferences dialog box.
2. Click the General icon in the Category list. The General settings appear. Click the Auto Answer check box.
3. Scroll down the Category list and click the LIVE Share icon. The LIVE Share setting appears. Click the Allow Remote Control box. The Password text field becomes active.
4. Enter your password, which can be up to eight characters long. As you type the password, only asterisks appear on the screen.
5. Click OK. You system is now ready for remote control. You can save power by turning off your monitor.

You can shut off remote control at any time by unchecking the Allow Remote Control setting in the LIVE Share preferences dialog box. You can also turn off automatic answering by unchecking the Auto Answer in the General preferences dialog box.

Beyond DVCs . . .

After our introduction to desktop video conferencing in Chapter 13 and our adventures with three specific systems in the succeeding chapters, we now round out our ISDN education with a variety of tips and resources gathered and organized for you in Part V that follows.

Part V
The Part of Tens

In This Part...

As its name implies, Part V provides you with tens upon tens of valuable resource nuggets accumulated during the writing of this book. Much of this information enhances and supports topics covered earlier. The following chapters provide extensive references to leading ISDN equipment vendors, ISDN Internet service providers, and other ISDN resources. You'll find contact information about ISDN equipment vendors for NT1 and NT1 Plus devices, ISDN telephones, remote access products, and desktop video conferencing systems. Other resources include ISDN resources available on the Internet and ISDN-related publications. Lastly, you'll find information about ways to save money getting ISDN and receive a bird's eye view of the steps needed to get up and running with ISDN.

Chapter 17

Ten Ways to Save Money Getting ISDN

In This Chapter

- Be a smart consumer when ordering your ISDN service from the telephone company
- Preserve your CPE investment in a constantly changing environment
- Do your own ISDN wiring and save a bundle
- Ask for promotional prices before you buy

ISDN can get expensive by the time you put all the pieces together, but there are ways to save money. Here are 10 ways you can avoid spending money unnecessarily.

Use the Caveat Emptor Defense

ISDN is a new technology, and there is a shortage of people who know what ISDN options are available and which ones are best for you. There are even fewer people who understand the needs of individuals and small businesses who are trying to get up and running with ISDN. Your only defense is to educate yourself as a consumer, even if you plan to have other people do the actual implementation. What you don't know can cost you time and money.

Know What You Need before Ordering ISDN Service

Get all the ISDN line provisioning information for all the devices you plan to use on your ISDN line before you order service. Telephone companies typically charge for making any changes to an ISDN line configuration after you place the initial order. If you will use an analog telephone connected to an NT1 Plus device or an ISDN telephone, make sure you know what calling services you want. Do it correctly from the start and you'll save an order charge ranging from $15 to $40 for each change.

Do Your Own Wiring

Wiring used for ISDN is the same as for POTS. Like analog service, your local telephone company brings the line to your premises. From there, it's your responsibility for the inside wiring. If yours is a simple inside wiring job, in many cases you can do it yourself and save the cost of telephone company or independent installers. A typical telephone company installer costs $50 to $75 per hour.

Don't Get an RJ-45 U-Interface Jack

You don't need an RJ-45 jack to terminate your U-interface. Most telephone companies and their installers don't tell you this fact. The result is that they install an RJ-45 jack for which they charge you far more than its cost. You typically pay about $15. As a bonus, you'll pay the telephone company to install it. You can buy an RJ-11 wall jack for $2 and install it yourself.

Shop Around for Your ISDN CPE

Many ISDN CPE vendors sell their products through the telephone companies at better prices than they themselves offer. Check with your telephone company or Internet service provider before buying directly from a CPE vendor. Typically, the manufacturer sells at list but offers price breaks if the equipment is sold through the telephone company at the time you order ISDN service. If you are buying multiple CPE units, ask for volume discounts.

Compare ISDN or Analog for Voice

Many telephone companies charge you more for using voice over ISDN than voice via analog. For example, NYNEX charges for ISDN calls made within a local calling area that, for residential analog service, would be included in the base rate. You may find it's cheaper to keep your analog line for outgoing voice communications and use your ISDN line for data transmission or for incoming calls or faxes.

Shop for ISDN Internet Service Providers

ISDN service for connecting to the Internet is new, and prices for service vary. The most progressive Internet service providers, such as InterNex and PSI, offer basic ISDN service for about $199 to start up and $29 per month thereafter. For the $29 per month, you receive 20-30 hours of free connect time. Keep in mind that these charges don't include any ISDN toll charges from your local telephone company. Try to find an Internet service provider that offers a telephone access number that is a local call. Internet service providers offering ISDN service often sell a package that includes the hardware for your end of the connection. Check out these packages, because they may offer a better deal than buying the product from the manufacturer.

Buy ISDN Products with a 30-Day Return Policy

ISDN is a complex realm in which things don't always work out as you expect. When you buy a product, use your credit card and make sure the vendor offers a 30-day return policy. The last thing you want is an expensive piece of ISDN equipment that doesn't fit your needs.

Ask about Telco Promotional Programs

Local and long-distance telephone companies are known to run promotional programs and not tell anyone. You need to ask — maybe even demand — to know. For example, AT&T ran a promotional program for 50% off its basic long distance ISDN service charges. NYNEX offered a promotion to waive the $190 installation charge for ISDN service. Even if the program expired, if it took you time just to get your service, demand the promotional price — even if they say it has expired.

Don't Get Locked into Restrictive CPE

Don't get locked into the wrong ISDN CPE products that restrict your options. For example, buying an ISDN adapter card with built-in NT1 for remote access without any S/T-interface ports will restrict you from using your PC for other applications, such as video conferencing. Always try to get ISDN CPE options that allow you the most flexibility so you can add other ISDN applications later.

Chapter 18

Ten Plus Five NT1, NT1 Plus, and ISDN Telephone Products

In This Chapter

- Leading NT1 products and vendors
- Leading NT1 Plus products and vendors
- Leading ISDN telephone vendors

The NT1 or NT1 Plus device is a pivotal piece of your ISDN connection. Here are the leading NT1 and NT1 Plus vendors as well as leading ISDN telephone vendors.

Where to Buy NT1, NT1 Plus, and ISDN Telephone Products

Each vendor sells products differently. In most cases you can order products directly from the vendor. However, some vendors only sell through resellers or telephone companies.

Try calling the vendor first. For vendors that only sell through resellers, check your local Yellow Pages under Telephone. Be aware that buying a product directly from the vendor may be more expensive than buying a product through a reseller. Chapter 19 provides a listing of the few ISDN CPE resellers.

NT1 Devices

NT1 devices provide the network termination function. They typically include two S/T-interface ports. Here are the leading products.

NT1 Ace
AdTran
901 Explorer Boulevard
Huntsville, AL 35806
Voice: (800) 971-8090
Fax: (205) 971-8699

Ask for AdTran's new NT1 product that is less expensive than the $395 list price for the NT1 Ace.

NT1U-100TC
NT1U-220TC
Tone Commander Systems, Inc.
4370 150th NE
Redmond, WA 98073
Voice: (800) 524-0024
Fax: (206) 881-7179

The NT1U-100TC device includes only one S/T-interface port and no PS2 power for ISDN devices. The NT1U-220TC is the standard two S/T-interface ports model. The great feature of the NT1U-220TC is that it has a reset button — the only NT1 device that does.

NT1D
Motorola
Transmission Products Division
500 Bradford Drive
Huntsville, AL 35805
Voice: (800) 451-2369

UT620
Alpha Telecom, Inc.
7501 South Memorial Parkway, Suite 212
Huntsville, AL 35802
Voice: (205) 881-8743
Fax: (205) 880-9720

AT&T's NT1U200
AT&T's NT1s aren't sold directly through AT&T. You must purchase them through an authorized AT&T reseller. Check with your telephone company, or look in your Yellow Pages under Telephones for AT&T resellers.

E-Tech NT1U700
E-Tech Research Inc.
1800 Wyatt Drive
Santa Clara, CA 95051
Voice: (408) 988-8108
Fax: (408) 988-8109

Northern Telecom NT1
Northern Telecom NT1s are not sold directly through Northern Telecom. You must purchase them through an authorized Northern Telecom reseller. Check with your telephone company, or look in your Yellow Pages under Telephones.

NT1 Plus Devices

Unfortunately, there are only a few vendors selling NT1 Plus devices, which are defined as including both RJ-11 analog ports and RJ-45 S/T-interface ports. Fortunately, the IBM and ADAK products are good choices for traveling the NT1 Plus route. The Super NT1 from Alpha Telecom was not available as of this book's writing, but it will include two RJ-11 analog ports and two RJ-45 S/T-interface ports.

7845 ISDN Network Terminator Extended
IBM Direct
4111 Northside Parkway
Atlanta, GA 30327
Voice: (800) 426-2255
Faxback: (800) 426-4329

ADAK 221
ADAK 421
ADAK Communications Corporation
5840 Enterprise Drive
Lansing, MI 48911
Voice: (517) 882-5191
Fax: (517) 882-3194

Super NT1
Alpha Telecom, Inc.
7501 South Memorial Parkway, Suite 212
Huntsville, AL 35802
Voice: (205) 881-8743
Fax: (205) 880-9720

ISDN Telephones

ISDN telephones are just now coming online at prices that make them attractive. Using an ISDN telephone lets you harness the powerful voice capabilities of ISDN.

Model 40d
Tone Commander Systems, Inc.
4370 150th NE
Redmond, WA 98073
Voice: (800) 524-0024
Fax: (206) 881-7179

LTI-1001LS
LTI-1501
LTI-2002
Lodestar Technology, Inc.
3101 Maguire Boulevard
Orlando, FL 32803
Voice: (407) 895-0881
Fax: (407) 895-3566

Lodestar currently offers the best deals in ISDN telephones. These telephones support National ISDN, AT&T 5ESS, and DMS100 custom switches. Lodestar ISDN telephones are sold under AT&T and Northern Telecom labels. The only distributor of Lodestar telephones is Cameron Communication Group, which is listed in Chapter 19 under ISDN CPE Resellers.

LTI1001SN
LTI1501SN
LTI2002SN
Northern Telecom, Inc.
4001 E. Chapel Hill Nelson Hwy.
Research Triangle Park, NC 27709
Voice: (800) 667-8437

These ISDN telephones, sold by authorized Northern Telecom resellers, are the Lodestar phones with a Northern label.

Optiset Model 787
Siemens Stromberg-Carlson
900 Broken Sound Parkway
Boca Raton, FL 33467
Voice: (407) 955-5000

Siemens' Optiset is a new ISDN telephone designed specifically for the home and small business market. It offers a modular approach that lets you expand its capabilities. These ISDN telephones support only National ISDN.

AT&T 8510
AT&T NI14
AT&T NI35

These ISDN telephones are sold via AT&T authorized resellers. Check with your telephone company or your Yellow Pages under Telephones.

Chapter 19

Ten Plus Four ISDN User Resources

In This Chapter

- ISDN resources on the Internet
- Books, newsletters, and other printed resources
- ISDN user groups
- ISDN equipment resellers

As an ISDN user, you need all the help you can get. Here is a collection of resources available to help on your journey toward ISDN enlightenment.

On the Internet

The two leading sources of ISDN information on the Internet are a World Wide Web site and an ISDN newsgroup. Both are outstanding resources that offer the most up-to-date information.

Dan Kegel's ISDN Page

The most comprehensive links to ISDN resources on the Internet are at a Web site maintained by Dan Kegel. This site provides links to all kinds of ISDN information, including equipment vendors, service providers offering ISDN connections, and much more. It's constantly updated with new information. This is truly a one-stop shopping site for any resources available on the Internet that pertain to ISDN. The online address for this site is:

```
http://alumni.caltech.edu/~dank/isdn/
```

Dan Kegel's ISDN Page includes more than 30 main headings and a few hundred links. Just to give you an idea of what's covered at this site, here are some of the entries from Dan Kegel's ISDN Page's table of contents.

ISDN Dialtone Providers
ISDN Internet Connectivity Providers
ISDN Interoperability Test Reports
ISDN Shareware/Free Software
Commercial ISDN Software
BRI (128 Kbps) ISDN ISA or MCA Bus Cards
Network Terminators for US BRI (NT1s)
ISDN Videoconferencing
ISDN Bridges and Routers with Ethernet Interfaces
ISDN Computer Networking Hardware Vendors
ISDN User Groups
The ISDN Newsgroup

The Usenet news includes an ISDN newsgroup that's a good source of information. Several ISDN gurus regularly participate in this newsgroup. It's a great resource for troubleshooting problems and keeping abreast of what's happening in the ISDN realm. Figure 19-1 shows a typical listing of articles in this newsgroup. The ISDN newsgroup is:

```
comp.dcom.isdn
```

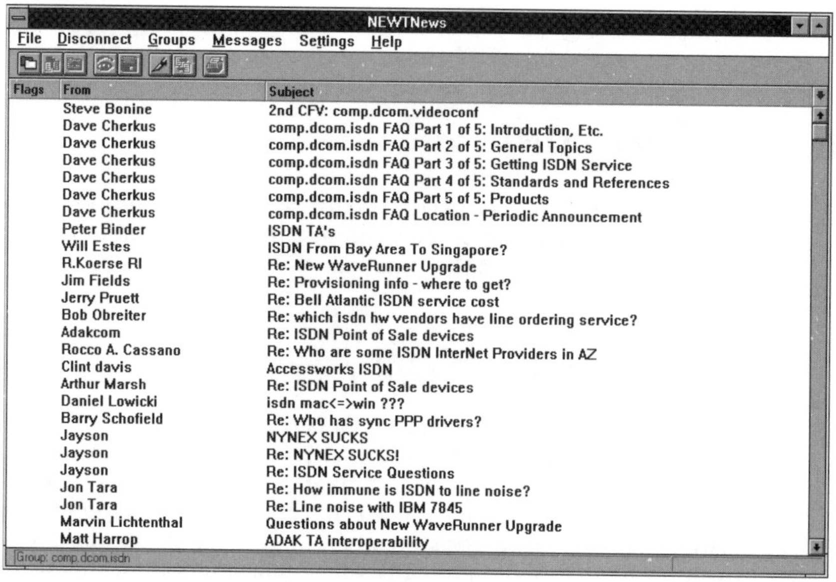

NEWTNews

File Disconnect Groups Messages Settings Help

Flags	From	Subject
	Steve Bonine	2nd CFV: comp.dcom.videoconf
	Dave Cherkus	comp.dcom.isdn FAQ Part 1 of 5: Introduction, Etc.
	Dave Cherkus	comp.dcom.isdn FAQ Part 2 of 5: General Topics
	Dave Cherkus	comp.dcom.isdn FAQ Part 3 of 5: Getting ISDN Service
	Dave Cherkus	comp.dcom.isdn FAQ Part 4 of 5: Standards and References
	Dave Cherkus	comp.dcom.isdn FAQ Part 5 of 5: Products
	Dave Cherkus	comp.dcom.isdn FAQ Location - Periodic Announcement
	Peter Binder	ISDN TA's
	Will Estes	ISDN From Bay Area To Singapore?
	R.Koerse RI	Re: New WaveRunner Upgrade
	Jim Fields	Re: Provisioning info - where to get?
	Jerry Pruett	Re: Bell Atlantic ISDN service cost
	Bob Obreiter	Re: which isdn hw vendors have line ordering service?
	Adakcom	Re: ISDN Point of Sale devices
	Rocco A. Cassano	Re: Who are some ISDN InterNet Providers in AZ
	Clint davis	Accessworks ISDN
	Arthur Marsh	Re: ISDN Point of Sale devices
	Daniel Lowicki	isdn mac<=>win ???
	Barry Schofield	Re: Who has sync PPP drivers?
	Jayson	NYNEX SUCKS
	Jayson	Re: NYNEX SUCKS!
	Jayson	Re: ISDN Service Questions
	Jon Tara	Re: How immune is ISDN to line noise?
	Jon Tara	Re: Line noise with IBM 7845
	Marvin Lichtenthal	Questions about New WaveRunner Upgrade
	Matt Harrop	ADAK TA interoperability

Group: comp.dcom.isdn

Figure 19-1: The comp.dcom.isdn newsgroup should be a regular stop as you learn to master ISDN.

ISDN Publications

Printed material about ISDN is scarce. However, there are a few publications worth exploring for the technical side of ISDN and for updates on what's happening in ISDN.

- *ISDN Second Edition*, Gary C. Kessler, McGraw-Hill, 1993. This is a technical textbook that provides the most detail on ISDN among books on the topic.
- *The ISDN Literacy Book*, Gerald L. Hopkins, Addison-Wesley, 1995. This book reduces the technical information to a more manageable level than the Kessler book. It also deals with the broader ISDN issues.
- *ISDN in Perspective*, Fred Goldstein, Addison-Wesley, 1992. Another technical book about ISDN, but the author does a reasonable job explaining technical topics given its age.
- *The Basic Book of ISDN*, 2nd Edition, Motorola University Press, 1992. An introduction to ISDN, but with a technical orientation. This 45-page book explains the essential concepts behind ISDN.
- *ISDN: A User's Guide to Services, Applications & Resources in California*, Pacific Bell, 1994. This 54-page book provides a very basic explanation of ISDN. It is available free by calling (800) 4PB-ISDN.
- *ISDN Solutions* is a slick magazine by the Corporation for Open Systems International that provides a solid resource about ISDN applications as well as other information. A single copy costs $20. To order, call (703) 205-2700.
- *The ISDN User Newsletter* offers bimonthly updates about ISDN happenings. It is published by Information Gatekeepers and costs $35 per year. Call (800) 323-1088 or (617) 232-3111 for more information.

ISDN User Groups

ISDN is so new that there are currently only a few user groups.

North American ISDN User's Forum (NIUF)
National Institute of Standards and Technology
Building 223, Room B364
Gaithersburg, MD 20899
Voice: (301) 975-2937

This is a government-sponsored organization made up mostly of ISDN industry people. It offers several useful publications, such as an ISDN wiring guide, provided as Appendix B.

California ISDN User's Group
P.O. Box 27901-318
San Francisco, CA 94127
Voice: (415) 241-9943
Fax: (415) 753-6942

This is by far the most active ISDN user group in the country. It publishes a newsletter and regularly holds events for ISDN users to sample ISDN CPE products and services.

ISDN CPE Resellers

There are few ISDN equipment resellers. Here are two that currently sell ISDN products to end users.

Bell Atlantic Teleproducts
West Building, Suite 150
50 E. Swedesford Road
Frazer, PA 19355
Voice: (800) 221-0845

The Bell Atlantic Teleproducts center sells a variety of ISDN CPE products.

Data Comm Warehouse
1720 Oak Street
P.O. Box 301
Lakewood, NJ 08701
Voice: (800) 328-2261
Fax: (908) 363-4823

Data Comm Warehouse sells mail order ISDN and other networking products at discount prices. For example, it sells AccessWorks Remote, Ascend Pipeline 50 and 50 HX, and the Gandalf 5242I ISDN Telecommuter Bridge. It has a limited line of ISDN products, but will probably add more as the demand for ISDN grows. Call for their free catalog.

Cameron Communications Group
1728 Carriage Court
Brentwood, TN 37027
Voice: (615) 370-0029
Fax: (615) 370-5855

This small distributor sells ISDN equipment and is the only independent distributor of Lodestar ISDN telephones.

Chapter 20

Ten Desktop Video Conferencing Systems and Resources

In This Chapter

- Leading desktop video conferencing system vendors
- Obtaining video conferencing via mail order
- Books on video conferencing

Video conferencing is one of the most exciting ISDN applications. Thanks to standards, it's also one of the most stable in terms of interoperability. Here are the leading vendors of video conferencing systems and related resources.

Video Conferencing Systems

The following desktop video conferencing systems represent the most affordable systems available. They all comply with the H.320 standard.

Vistium Personal Video 1200
AT&T Global Information Solutions
1700 S. Patterson Boulevard
Dayton, OH 45479
Voice: (800) 447-1124

AT&T doesn't sell the Vistium 1200 directly. Call them for referral to an authorized reseller in your area.

PictureTel Live PCS 50
PictureTel Corporation
The Tower at Northwoods
222 Rosewood Drive
Danvers, MA 01923
Voice: (800) 716-6000 or (508) 762-5000
Fax: (508) 762-5245

ProShare Personal Video Conferencing System 200
Intel Corporation
2200 Mission College Boulevard
Santa Clara, CA 95052
Voice: (800) 538-3373 or (503) 629-7354
Fax: (800) 525-3019

Vivo320
Vivo Software, Inc.
411 Waverly Oaks Road
Waltham, MA 02154
Voice (617) 899-8900
Fax: (617) 899-1400

InVision for Windows
InVision
317 S. Main Mall, Suite 310
Tulsa, OK 74103
Voice: (918) 584-7772
Fax: (918) 584-7775

InVision is a software-only desktop video conferencing product that works via an Ethernet or a TCP/IP network. It supports a number of third-party video capture cards.

Communicator III
EYETEL Technologies, Inc.
267 W. Esplanade, Suite 206
North Vancouver, BC
Canada
Voice: (604) 984-2522
Fax: (604) 984-3566

The Communicator III is an expensive but high-quality desktop video conferencing system.

PicturePhone Direct

PicturePhone Direct provides a one-stop solution for video conferencing systems and a host of related products through mail order. Its product line includes camcorders, slide-to-video systems, headsets, tripods, NT1s, and books.

PicturePhone Direct
200 Commerce Drive
Rochester, NY 14623
Voice: (800) 521-5454
Fax: (716) 359-4999
E-mail: PPP000231@interamp.com
Web: http://ppd.gems.com/ppd

Videoconferencing Books

The following two books are available from PicturePhone Direct.

- *VideoConferencing Secrets* by Jeremy Goldstein teaches you how to communicate effectively via video conferencing. It explains how to use visual materials and equipment for video meetings, how to use visual materials to help reduce meeting time, how to use secrets from television broadcasting to help your video image, and how to create excitement and maintain interest in your video presentation.
- *Videoconferencing and Money, Money, Money* by Jeremy Goldstein is a good primer on video conferencing. It covers applications for video conferencing, the benefits of video conferencing, how to buy desktop video conferencing equipment, and how video conferencing works. It also provides solid information about conducting a cost-benefit analysis of video conferencing compared to the alternative of travel, as well as about productivity improvements.

AT&T WorldWorx

WorldWorx personal conferencing services assist video conference users to set up multiuser video conferences. Using WorldWorx, you can schedule a video conference for as many as 22 people. You set up a video conference by providing participant telephone numbers to AT&T; they do the rest. To find out more about AT&T WorldWorx, call (800) 828-WORX (828-9679).

Chapter 21

Ten ISDN Internet Surfing Resources

In This Chapter

- Internet service providers that offer dialup ISDN service
- Places to find new ISDN Internet service providers
- Leading ISDN TCP/IP programs for surfing the Internet

This chapter covers resources that will help you as you use ISDN to surf the Internet.

ISDN Internet Service Providers

Currently, only a few Internet service providers offer ISDN dialup services at affordable prices — but this situation is quickly changing. The popularity of the World Wide Web and the demands it makes for more bandwidth are the leading pulls for ISDN connections to the Internet. Expect to see more Internet service providers offering ISDN service.

InterNex Information Services
1050 Chestnut Street, Suite 202
Menlo Park, CA 94025
Voice: (415) 473-3060
Fax: (415) 473-3062

InterNex provides affordable ISDN connections for Northern California, and they're planning to go national.

CONNECTnet
6370 Lusk Boulevard, Suite F208
San Diego, CA 92121
Voice: (619) 450-0254
Fax: (619) 450-3216
E-mail: george@connectnet.com

Internet Exchange, Ltd.
5 Commonwealth Road
Natick, MA 01760
Voice: (508) 647-4726
Fax: (508) 647-4727
E-mail: office@ixl.net

TerraNet, Inc.
729 Boylston Street
Boston, MA 02116
Voice: (617) 450-9000
Fax: (617) 450-9003
E-mail: sales@terra.net

QuakeNet
830 Wilmington Road
San Mateo, CA 94402
Voice: (415) 655-6607
Fax: (415) 377-0635
E-mail: admin@Quake.Net

CERFnet
P.O. Box 85608
San Diego, CA 92186
Voice: (800) 876-2373
Fax: (619) 455-3990
E-mail: sales@cerf.net

This leading Internet service provider offers dialup ISDN service via a toll-free number. For individuals and small businesses, this approach can be economical. It's worth checking out.

PSI

PSI is the leading Internet service provider offering ISDN service. Its InterRamp service is available nationwide via local telephone numbers. The InterRamp service works with the ISDN*tek card or 3Com's Impact digital modem and NetManage's Internet Chameleon. Table 21-1 lists InterRamp ISDN access numbers.

Performance Systems International (PSI)
510 Huntmar Park Drive
Herndon, VA 12180
Voice: (800) 827-7482 or (703) 620-6551

Table 21-1: PSI InterRamp ISDN access numbers by state and city

State and City	*ISDN Number*
Arkansas	
Little Rock	(501) 340-6400
Arizona	
Phoenix	(602) 640-6096
Tucson	(520) 620-6152
California	
Burbank	(818) 566-7200
Fresno	(209) 497-0200
Los Angeles	(213) 623-6411
Orinda	(510) 253-9500
San Ramon	(510) 244-1500
San Diego	(619) 230-1221
Santa Ana	(714) 285-9100
Sacramento	(916) 537-2901
Santa Clara	(408) 289-1510
San Francisco	(415) 357-9900
	(415) 442-4600
Sunnyvale	(415) 390-0900
Van Nuys	(818) 781-3800
Colorado	
Colorado Springs	(719) 594-2090
Connecticut	
Hartford	(203) 947-7656
District of Columbia	
Washington	(202) 408-3199
Delaware	
Wilmington	(302) 576-0500
Florida	
Jacksonville	(904) 355-1523
Miami	(305) 470-6277
Orlando	(407) 648-9426
Tampa	(813) 276-9824

(continued)

State and City	ISDN Number
Georgia	
Atlanta	(404) 865-0110
Illinois	
Chicago	(312) 565-9446
Urbana	(217) 337-3020
Indiana	
Fort Wayne	(219) 452-1004
Indianapolis	(317) 576-0308
South Bend	(219) 271-0326
Massachusetts	
Boston	(617) 450-6800
Maryland	
Baltimore	(410) 244-0719
Maine	
Portland	(207) 791-8300
Michigan	
Battle Creek	(616) 226-8589
Bay City	(517) 667-6121
Birmingham	(810) 574-0259
Detroit	(313) 225-4099
Flint	(810) 768-7980
Grand Rapids	(616) 235-9246
Kalamazoo	(616) 226-8593
Lansing	(517) 336-8936
Midland	(517) 837-1394
Pontiac	(810) 475-4425
Missouri	
Kansas City	(816) 235-7000
St. Louis	(314) 516-0000
North Carolina	
Research Triangle Park	(919) 558-2121
New Jersey	
Newark	(201) 622-6100
New Mexico	
Albuquerque	(505) 246-9088
Nevada	
Las Vegas	(702) 382-6667

(continued)

State and City	ISDN Number
New York	
Albany	(518) 436-3200
Islip	(516) 468-5100
New York	(212) 560-4560
	(212) 631-2860
Rochester	(716) 324-1000
Syracuse	(315) 448-4000
White Plains	(914) 993-6000
Ohio	
Akron	(216) 374-6990
Cincinnati	(513) 792-9229
Cleveland	(216) 696-8236
Columbus	(614) 222-0825
Dublin	(614) 228-7452
Oklahoma	
Tulsa	(918) 445-2622
Oklahoma City	(405) 290-5000
Oregon	
Portland	(503) 220-4086
Pennsylvania	
Harrisburg	(717) 720-0528
Philadelphia	(215) 587-9450
Pittsburgh	(412) 562-1103
Texas	
Abilene	(915) 738-3100
Austin	(512) 432-0001
Dallas	(214) 742-7907
	(214) 953-3199
El Paso	(915) 545-1400
Fort Worth	(817) 568-5800
Houston	(713) 567-0300
San Antonio	(210) 244-1900
Utah	
Salt Lake City	(801) 264-7940
Virginia	
Herndon	(703) 904-9050
	(703) 904-7129
Norfolk	(804) 455-8178
Richmond	(804) 755-7018
Washington	
Seattle	(206) 441-2203

Where to Check for New ISDN Internet Service Providers

A comprehensive listing of Internet service providers offering ISDN service is available at Dan Kegel's ISDN Page. The following World Wide Web page provides a constantly updated list of ISDN Internet service providers:

```
http://alumni.caltech.edu/~dank/isdn/isp.html
```

ISDN Internet Surfing Software

The leading TCP/IP application for surfing the Internet via ISDN is the Chameleon family of Windows TCP/IP software. Other TCP/IP software vendors will probably support ISDN in the near future.

ChameleonNFS (version 4.5 or later)
Internet Chameleon (version 4.1 or later)
NetManage
10725 North De Anza Boulevard
Cupertino, CA 95014
Voice: (408) 973-7171
Fax: (408) 257-6405
E-mail: info@netmanage.com
WWW: http://www.netmanage.com

ChameleonNFS is for networks, while Internet Chameleon is for single-user PCs. The ChameleonNFS works via Ethernet-based ISDN adapter card, or a stand-alone bridge and router. The Internet Chameleon works with ISDN adapter cards that support the WinISDN API. The only card that currently supports the WinISDN API is the ISDN*tek card, which is part of the PSI InterRamp package. The IBM WaveRunner will support the WinISDN API in the near future.

Internet-In-A-Box
Internet Office
Spry, Inc.
316 Occidental Avenue South, Suite 200
Seattle, WA 98104
Voice: (800) 777-9638
Fax: (206) 447-9008
E-mail: info@spry.com
WWW: http://www.spry.com

Chapter 22

More than Ten Remote Access Products

In This Chapter

- ISDN Ethernet-based ISDN adapter cards
- ISDN single-user ISDN bridges
- ISDN Ethernet-based, multiuser ISDN bridges
- ISDN serial communication cards
- Digital modems

The function of the remote access device is to allow your PC to connect to a host computer on a LAN, to the Internet, or to an online service provider. Here is a list of leading ISDN remote access devices.

Remote Access Hardware

A growing number of ISDN CPE vendors are getting into the remote access business. Among the several types of ISDN remote access devices are:

- Ethernet-based ISDN adapter cards
- Ethernet-based single-user, stand-alone bridges
- Ethernet-based stand-alone, multiuser ISDN bridges and routers
- Serial-based ISDN adapter cards
- Digital modems that support both asynchronous and synchronous communication

Any remote access device you plan to use should comply with PPP/MP (PPP Multilink Protocol), which allows you to surf the Internet using both B channels.

Ethernet-Based ISDN Adapter Cards

Ethernet-based ISDN adapter cards use Ethernet to communicate between your PC and ISDN. This hardware solution is less expensive for remote access than a stand-alone, single-user ISDN bridge.

DataFire
Digi International
6400 Flying Cloud Drive
Eden Prairie, MN 55344
Voice: (800) 344-4273 or (612) 943-9020
Fax: (612) 943-5398
WWW: http://www.digiboard.com

RemoteExpress ISDN LAN Adapter
RemoteExpress ISDN Bridge Pack
Intel Corporation
2200 Mission College Boulevard
P.O. Box 58119
Santa Clara, CA 95052
Voice: (800) 538-3373
Fax-back: (800) 458-6231
WWW: http://www.intel.com

EVERYWARE 1000
Combinet
333 West El Camino Real
Sunnyvale, CA 94087
Voice (800) 967-6651 or (408) 522-9020
Fax: (408) 732-5479
E-mail: info@combinet.com or sales@combinet.com
WWW: http://www.combinet.com

Stand-Alone, Single-User ISDN Bridges

Single-user ISDN bridges use an Ethernet card to connect a PC to a stand-alone bridge unit. The main advantage of the single-user, stand-alone unit is that you can connect any computer that supports Ethernet. One of the best Ethernet cards is the 3Com EtherLink III Combo, because it includes a built-in RJ-45 port to connect to an ISDN bridge.

EtherLink III Combo
3Com Corporation
5400 Bayfront Plaza
Santa Clara, CA 95052
Voice: (800) 638-3266

EVERYWARE 150
EVERYWARE 160
Combinet
333 West El Camino Real
Sunnyvale, CA 94087
Voice (800) 967-6651 or (408) 522-9020
Fax: (408) 732-5479
E-mail: info@combinet.com or sales@combinet.com
WWW: http://www.combinet.com

Pipeline 50 HX
Ascend Communications, Inc.
1275 Harbor Bay Parkway
Alameda, CA 94502
Voice: (800) 621-9578 or (510) 769-6001
Fax: (510) 814-2300
E-mail: info@ascend.com
WWW: http://www.internex.net/ascend/home.html

IMAC
Dual IMAC
Digi International
6400 Flying Cloud Drive
Eden Prairie, MN 55344
Voice: (800) 344-4273 or (612) 943-9020
Fax: (612) 943-5398
WWW: http://www.digiboard.com

Multiuser ISDN Bridges and Routers

Multiuser bridges and routers connect as a node on your network and allow you to route your LAN traffic to networks with multiple protocols. You can connect any computer that supports Ethernet-based networking.

Pipeline 50
Ascend Communications, Inc.
1275 Harbor Bay Parkway
Alameda, CA 94502
Voice: (800) 621-9578 or (510) 769-6001
Fax: (510) 814-2300
E-mail: info@ascend.com
WWW: http://www.internex.net/ascend/home.html

EVERYWARE 2000
Combinet
333 West El Camino Real
Sunnyvale, CA 94087
Voice (800) 967-6651 or (408) 522-9020
Fax: (408) 732-5479
E-mail: info@combinet.com or sales@combinet.com
WWW: http://www.combinet.com

LANLine 5242i
Gandalf Technologies, Inc.
Cherry Hill Industrial Center, Suite 9
Cherry Hill, NJ 08003
Voice: (800) 426-3253 or (609) 461-8100
Fax: (609) 461-5186

NELINK 1000
Network Express, Inc.
4251 Plymouth Road, Suite 1100
Ann Arbor, MI 48105
Voice: (313) 761-5005
Fax: (313) 995-1114

External Digital Modems

External digital modems connect to your PC via a serial port card. Digital modems typically perform asynchronous to synchronous conversion so you can connect to an Internet service provider using the high-speed capabilities of ISDN. External digital modems also include analog ports for telephones, faxes, and answering machines. The following are the leading vendors for external digital modems.

3Com Impact
3Com Corporation
5400 Bayfront Plaza
Santa Clara, CA 95052
Voice: (800) 638-3266

BitSURFR
Motorola
5000 Bradford Drive
Huntsville, AL 35805
Voice: (800) 365-6456 or (205) 430-8000
Fax: (205) 430-8926

ISU Express
ADTRAN
901 Explorer Boulevard
Huntsville, AL 35806
Voice: (205) 971-8000

ADAK Model 221
ADAK Model 421
ADAK Communications Corporation
5840 Enterprise Drive
Lansing, MI 48911
Voice: (517) 882-5191
Fax: (517) 882-3194

These products are NT1 Plus devices that also act as digital modems.

Serial ISDN Adapter Cards

The following serial communications adapter cards support the WinISDN API. So they work with Internet Chameleon, which is the nonnetworked version of the popular Microsoft Windows TCP/IP program.

Internet Card
Commuter Card
ISDN*tek
P.O. Box 3000
San Gregorio, CA 94074
Voice: (415) 712-3000
Fax: (415) 712-3003
WWW: http://isdntek.com

IBM WaveRunner
IBM Direct
4111 Northside Parkway
Atlanta, GA 30327
Voice: (800) 426-2255
Faxback: (800) 426-4329
WWW: http://www.ibm.com

Chapter 23

More than Ten Places to Get ISDN Service Answers

In This Chapter

- Where to get ISDN service information from local telephone companies
- Where to get ISDN service information from long-distance telephone companies
- Where to get ISDN service information from other sources

Getting information about ISDN service, such as availability and costs, can involve a variety of sources, including your local telephone company, long distance telephone companies, CPE vendors, and others. This chapter provides a listing of these sources.

Local Telephone Companies

Local telephone companies are your first source of ISDN service information. However, the quality of information from them varies widely from company to company. Table 23-1 lists the sources of ISDN information from local telephone companies in the United States.

Table 23-1: ISDN service information from local U.S. telephone companies

Telephone Company	*ISDN Information*
Ameritech	(800) 832-6328
Bell Atlantic	(800) 570-ISDN
New Jersey Bell	(800) 843-2255
Bell South	(800) 858-9413
Cincinnati Bell	(513) 566-DATA
Pacific Bell	Automated ISDN Availability Hotline: (800) 995-0346 ISDN Telemarketing (ordering): (800) 662-0735 ISDN Service Center: (800) 4PB-ISDN
GTE	Automated ISDN information service: (800) 4GTE-SW5 Florida, North Carolina, Virginia, and Kentucky: (800) 483-5200 Illinois, Indiana, Ohio, and Pennsylvania: (800) 483-5600 Oregon and Washington: (800) 483-5100 Hawaii and California: (800) 643-4411 Texas: (800) 483-5400
Nevada Bell	(702) 688-7124
NYNEX	Automated Availability System: (800) GET-ISDN Automated ISDN information system: (800) 438-4736, (800) 698-0817, or (212) 626-7297
Rochester Telephone	(716) 777-1234
Southwestern Bell	Austin: (512) 870-4064 Dallas: (214) 268-1403 Houston: (713) 567-4300 San Antonio: (512) 351-8050 St. Louis: (800) SWB-ISDN
US West	(303) 896-8370

Long Distance ISDN Service

If you plan to use your ISDN's circuit-switched data service outside your local telephone company's calling area, you must establish a long-distance service account with a long distance telephone company. This account is for circuit-switched data calls only. Normal voice telephone calls are included on your ISDN bill from the local telephone company.

Long distance calling charges for circuit-switched data calls arrive in a separate bill directly from the long distance company. Call the following numbers for more information:

- AT&T: (800) 222-7956
- MCI: (800) 766-2887
- Sprint: (800) 788-8981

CPE Vendors and Other Sources

ISDN equipment vendors are good sources for help about ISDN service because they deal with telephone companies on behalf of their customers. If you know what products you plan to use before getting ISDN service, contact the vendors to see if they can help you with ISDN service. Here are two leading sources of ISDN service information available to anyone.

- Intel provides ISDN service information at (800) 538-3373. They can tell you if ISDN is available in your area and can help you obtain ISDN service.
- Bellcore, the research organization for regional telephone companies, provides a national ISDN hotline at (800) 992-4736.

Chapter 24

Ten Steps to Digital Enlightenment via ISDN

In This Chapter

- Checking availability and CO switch type
- Determining your BRI configuration options and what your ISDN service will cost
- Getting wired for ISDN
- Developing your CPE game plan and equipping yourself with the right CPE
- Assembling your ISDN line provisioning information and ordering ISDN service
- Setting up your ISDN devices

Getting up and running with ISDN is a journey through a maze of interrelated steps. This chapter paints in broad brush strokes the steps needed to attain true digital enlightenment via ISDN. Its steps provide an overview of the key pieces involved in getting up and running with ISDN. Each of the ten steps references where you can find further information in this book.

Check Availability

The first order of business is to check the availability of ISDN service for your specific location. Most telephone companies offer automated systems for checking ISDN availability based on your area code and your prefix or exchange number (the first three numbers after your area code). Even if the automated system tells you ISDN service is available, confirm with a telephone company representative. There may be a limitation of 18,000 feet from your Central Office to your premises for ISDN service.

Chapter 3 covers how to check availability from your local telephone company, and Chapter 23 provides more information about obtaining answers to ISDN availability questions.

Find Out the CO Switch Type

If ISDN service is available in your area, find out the type of switch used by the telephone company for your ISDN service. Different switches have different capabilities for handling multiple ISDN devices and for other configuration options. There are two main switch platforms used by telephone companies, which are the AT&T 5ESS and NT DMS-100. These may use any of the following software options: AT&T 5ESS Custom, AT&T 5ESS NI-1 (National ISDN), and NT DMS 100 (National ISDN). Determining the type of switch at the CO helps you plan which ISDN applications and CPE configurations you can use.

Chapter 2 explains the function of telephone company switches in ISDN. Chapter 3 provides specific information about the role that telephone company switches play in establishing your ISDN service.

Determine Your BRI Configuration Options

While the standard for a BRI connection is two B channels and one D channel, telephone companies offer a number of configuration options. For example, you can get only one B channel or just two B channels without a D channel. But, in most cases you'll want the standard two B channels and one D channel option, or a D channel for signaling only. This latter service is referred to as 2B+0D.

You also must determine the channel configuration options that are available for your BRI line's B channels. The available channel configuration options are: circuit-switched voice (CSV) only, circuit-switched data (CSD) only, alternate voice/circuit-switched data (CSV/CSD), and packet data only. However, some telephone companies don't allow you to configure the B channels any way you want. For example, NYNEX doesn't allow you to configure both B channels as alternate voice/data.

Beyond figuring out channel configurations, you need to check a few other important pieces of information with your telephone company. Make sure your BRI connection supports multipoint configurations so you can work with more than one ISDN device at the same time. If you plan to use an analog telephone or ISDN telephone, find out what supplementary services for call management are available from your telephone company.

Chapter 2 explains the fundamentals of the Basic Rate Interface (BRI) and Chapter 3 covers specific BRI configuration issues. ■

Figure Out What ISDN Will Cost

Get the total cost from your telephone company to install ISDN and the monthly recurring costs. ISDN service costs include a one-time installation charge, recurring monthly charges, and usage charges. Your installation and recurring charges are affected by the configuration of your BRI line. Usage charges also vary, depending on whether the call is voice or data. Request ISDN tariffs from your local telephone company.

Chapter 3 covers ISDN service costs and Chapter 17 provides 10 ways to save money when you obtain and use ISDN. ■

Get Wired for ISDN

One of the first ISDN wiring issues to consider is whether you want to convert an existing POTS line to ISDN service or to add a completely new line. Whether your premises are a business site or a residence, and the location of the premises, impact which options you can choose. You can bring in a new line for ISDN, use an unused wire pair, or replace an existing analog line with ISDN service. If the telephone company can add a new line to your premises, you may want to do it. It takes time to climb the ISDN learning curve, so you may not want to hold your primary analog communication line hostage as you make the transition to ISDN. Additionally, you may want to use your ISDN line exclusively for data transmission and not tie it up with incoming voice calls.

Because ISDN uses the same wiring as POTS, you can do your own premises wiring from the demarcation point. If you don't want to do your own wiring, the telephone company installer can do it for $55 to $75 per hour. Remember, ISDN uses the same RJ-11 jack and connector as analog for the U-interface. However, ISDN devices use RJ-45 cabling and connectors to connect to the NT1 device.

Chapter 4 explains what is involved in doing your own wiring for ISDN. Chapter 17 provides tips for saving money in getting wired for ISDN. ■

Develop Your CPE Game Plan

Understanding the pieces is one thing, understanding how to assemble them into a CPE package to use your ISDN connection to its fullest is another. Assembling all the pieces for an effective ISDN connection requires some visualization of the layout options as defined by different applications.

An ISDN CPE game plan is essential to get the most from an ISDN line. The end result is that you want the flexibility to work with multiple applications on your ISDN line. You can use ISDN for several applications, including voice communications, remote access, and video conferencing.

The key piece of the puzzle to get the most from your ISDN connection is the placement of the NT1 function. In most cases, you'll want a stand-alone NT1 device or an NT1 Plus device. From this point you can add devices for voice communication, remote access, and video conferencing. If you want to connect an analog device to the line, you need an NT1 Plus device. If you plan to connect only ISDN-ready devices to your ISDN line, use an NT1 device. An ISDN-ready device can be your PC with an ISDN adapter card to prepare it for remote access or video conferencing.

Chapter 4 explains factors to include in developing your ISDN equipment implementation plan. ■

Equip Yourself with the Right ISDN CPE

ISDN equipment comes in two flavors, U and S/T, and CPE vendors typically sell their products in both flavors. ISDN equipment made for the U reference point means the NT1 functional device is built in. The significance of these two reference points is pivotal to getting the most from your ISDN connection. The bottom line is that you can't use multiple U-interface devices on the same ISDN line, because they include the NT1 function. ISDN equipment made for the S/T-interface requires the NT1 function to connect to ISDN. You can use multiple S/T-interface devices on an ISDN line, because they don't include built-in NT1s. In most cases, you should purchase only S/T-interface ISDN equipment that connects to a stand-alone NT1 or NT1 Plus device.

Chapter 4 starts you out with determining what equipment is right for you. From there, Part II (NT1s, NT1 Plus Devices, and ISDN Telephones), Part III (Remote Access via ISDN), and Part IV (Face to Face via Desktop Video Conferencing), provide you with the specifics about ISDN CPE. ■

Assemble Your CPE Provisioning Information

Each application and its associated device require special configurations about which you need to tell the telephone company when ordering ISDN service. Most CPE products include that provisioning information in their documentation. To prepare to order ISDN service, you need to pull together the ISDN provisioning information for all devices you plan to connect to your ISDN line. The telephone company will take this information to program its switch at the CO to work with your equipment at the time you order ISDN service.

Chapter 3 covers the essentials to provision your ISDN connection and Chapter 4 goes into more specifics about provisioning your ISDN line for multiple uses. ■

Order Your ISDN Service

Ordering your ISDN line connection involves exchanging information with the telephone company. This information includes the specifics about what you need for your BRI line and CPE. In turn, from the telephone company you receive information to configure your CPE to work with the ISDN connection.

You telephone company may assign you SPIDs for each device connected to the ISDN line, depending on the type of switch at the CO. If the telephone company assigns you SPIDs, you'll need them to configure your ISDN devices. The telephone company also assigns you one or more directory numbers, depending on your ISDN configuration.

After you establish ISDN service from your local telephone company, you establish an account with a long distance telephone company for circuit-switched data. Unlike analog service, the local telephone company doesn't automatically establish your long distance service for circuit-switched data at the time you order your local service. After you establish an account from a long distance telephone company, you'll receive a separate monthly bill for circuit-switched data from that company.

Chapter 3 explains the factors involved in ordering your ISDN service, and Chapter 4 builds on these fundamentals by telling you how your choice of applications affects your choice of service. ■

Set Up Your ISDN Devices

At last you're beginning to see light at the end of the digital tunnel! The final step is to decide where you install your ISDN hardware and software. To set up your ISDN devices, you'll need the information from your telephone company, such as SPIDs, directory numbers, and switch type. Typically, the first device to get up and running is your NT1 or NT1 Plus device. Once it's running, you can set up your other ISDN devices. Getting all your devices to work takes time and patience — stay cool and calm. The results are well worth all your effort!

Part II (NT1s, NT1 Plus Devices, and ISDN Telephones), Part III (Remote Access via ISDN), and Part IV (Face to Face via Desktop Video Conferencing) cover the specifics about setting up specific CPE for ISDN. ■

On to the Digital Light

You've reached the end of your journey to digital enlightenment via ISDN. Along the way you laughed, you cried, but your patience and perseverance has paid off. Now, you're in the driver's seat as you take control of your digital communication future. Your work and play will never be the same again, thanks to ISDN. You also have a solid foundation for continuing your journey to new and even greater ISDN adventures. . . .

Appendix A

Into the ISDN Tariff Maze

In This Appendix

- Understanding the Zen of ISDN service tariffs
- Basic ISDN tariffs for the seven RBOCs

Determining the total cost of ISDN service from your local telephone company can become the search for the Holy Grail. Only a few telephone companies have boiled down the complex pricing for ISDN service into a friendly price list. This appendix provides you with some basic ISDN tariff information for the seven RBOCs.

Tariff Terrorism

In the real world, determining the price of a product or service is typically as easy as looking or asking. But the telephone companies operate in a regulated, monopolistic environment that isn't consumer driven. ISDN pricing, called tariffs in telephone company lingo, comes together through the application of complex cost allocation and recovery rules established by both federal and state regulators. Thus, ISDN pricing varies from one telephone company to another and from one state to another.

Don't take any of the prices in this appendix as gospel. Many of these prices were gleaned from complex tariff data that defy logic. In addition, ISDN prices and services are constantly changing, so you need to contact your local telephone company. Chapter 23 provides a list of local telephone company phone numbers to get ISDN information. ■

Determining your costs and options is one of the most difficult things about getting ISDN service. Getting ISDN service prices from some telephone companies can be a real nightmare. Even when you get the information, be prepared to spend some time determining what it really will cost you. Many

RBOCs present a smorgasbord of configuration options) with each one typically having its own installation and monthly charges. In some cases you may never really know the true cost of your ISDN service until you get the bill.

When you contact your telephone company to find out about ISDN service and costs, ask for any ISDN promotional programs that can save you money. ■

Bellcore's ISDN Tariff Summary

Bellcore generates a handy summary report of ISDN tariffs for the RBOCs, but unfortunately the last issue was June, 1994. However, their report is still useful for estimating costs for ISDN service. They plan to generate a new ISDN Tariff Summary in the near future. Call the Bellcore hotline at (800) 992-4736 for more information.

Ameritech

Ameritech services Illinois, Indiana, Michigan, Ohio, and Wisconsin. Table A-1 lists estimated costs for a 2B+D ISDN configuration with two B alternate voice/circuit-switched data channels and a D channel. This data comes from Bellcore's Tariff Summary for June 1994. For more detailed information, call Ameritech at (800) 832-6328, and they'll fax you an ISDN Direct Price List.

Table A-1: Ameritech's ISDN pricing

State	*Installation Charges*	*Monthly Charges*
Illinois	$179.50	$39.15
Indiana	$154.00	$97.00
Michigan	$162.00	$40.21
Ohio	$144.35	$43.60
Wisconsin	$159.65	$40.00

The FCC needs to see the digital light

The Federal Communications Commission (FCC) now charges you a $3 a month for each telephone line. These subscriber line charges — called SLCs — are collected by your local telephone company. Your long distance telephone company also collects surcharges for the FCC via your long distance toll calls. The FCC then redistributes the money back to the local telephone companies. The FCC has been redistributing SLCs since the breakup of AT&T. Now the FCC has defined an ISDN line having two telephone lines--one line for each B channel. The result of this move is a doubling of the SLC to $6 a month. As a result, users can expect to see 50 to 100 percent increases in ISDN line costs. If the FCC sticks to its arrogant position, the only digital communications option available to the rest of us may become out of reach.

Bell Atlantic

Bell Atlantic services Delaware, Maryland, New Jersey, Pennsylvania, Virginia, Washington, DC, and West Virginia. Table A-2 lists the tariffs for 2B+D configurations. If you request any additional work beyond the Network Interface, a $42.00 service call charge and $16.00 for every 15 minutes of labor will apply. Table A-3 lists local area calling plans for circuit-switched data.

Table A-2: Bell Atlantic ISDN service charges

State	*One-Time Installation*	*Monthly 2B+D Service*	*Monthly Business Line Service*
Delaware	$141.50	$19.50	$31.95
Maryland	$131.50	$19.50	$19.26
New Jersey	$142.25	$19.50	$19.65
Pennsylvania	$169.75	$19.50	$28.50
Virginia	$100.00	$19.50	$19.34
Washington, DC	$120.75	$19.50	$16.77
West Virginia	$136.40	$19.50	$27.25

Table A-3: Bell Atlantic circuit-switched data local charges

Data Usage Option (Local Calls Only)	*Monthly Allowance*	*Monthly Rate*
Option 1	None	5 cents per minute per B channel
Option 2	25 hours per month	$60.00, plus 5 cents per minute beyond 25 hours per B channel
Option 3	50 hours per month	$90.00, plus 5 cents per minute beyond 50 hours per B channel
Option 4	100 hours per month	$120.00, plus 2 cents per minute beyond 100 hours per B channel

Bell South

The price range for Table A-4 is based on four different 2B+D configuration options, which Bell South refers to as Capability Packages. The following describes these four Capability Packages.

FYI

You'll know what ISDN costs only after you get the bill

In the case of my own experience with NYNEX, I didn't know what ISDN would cost until I got the bill. While NYNEX does publish a tariff, getting it was difficult. And figuring it out was even more difficult. Even when I thought I had a rough estimate, it still cost more. My cost for ISDN service was around $197 for installation. That $197 installation charge was after they knocked off $190 for a special promotion. So the cost would have been $387. The basic ISDN service runs around $43. This price reflects a reduction of $32 resulting from an ISDN promotional program. After the first six months of service, the rate goes to $75. This $32 per month charge is for NYNEX's Virtual ISDN Service, which is the routing of your ISDN line to a CO with a switch that can handle ISDN. NYNEX charges you for everything, including the privilege of having a telephone number.

- Capability Package N includes alternate voice/circuit-switched data on one B channel, circuit-switched data on the other B channel, and basic D channel packet-switched data. This package provides non-EKTS voice features including Flexible Calling, Additional Call Offering, and Calling Number Identification. Data capabilities include Calling Number Identification.
- Capability Package O is equivalent to Capability Package N, but with CACH EKTS voice service. This package provides more Call Offering functionality.
- Capability Package P includes alternate voice/circuit-switched data on two B channels and basic D channel packet-switched data. This package includes the same non-EKTS voice features as Capability Package N.
- Capability Package Q is equivalent to Capability Package P, but with CACH EKTS voice service. This package provides more Call Offering functionality.

Table A-4: Bell South ISDN tariffs for residential and business customers

State (R)Residential (B)Business	*Installation Charges*	*Monthly Charges*
Alabama	$221.75-$228.75 (R) $253.75-$267.75 (B)	$72.35-$79.35 (R) $111.50-$125.50 (B)
Florida	$211.00-$218.00 (R) $230.00-$244.00 (B)	$60.65-$67.65 (R) $111.50-$125.50 (B)
Georgia	$202.50-$209.50 (R) $229.25-$243.25 (B)	$66.90-$73.90 (R) $111.50-$125.50 (B)
Kentucky	$254.10-$261.10 (R) $274.11-$288.10 (B)	$64.55-$71.55 (R) $111.50-$125.50 (B)
Louisiana	$267.11-$274.11 (R) $270.11-$284.11 (B)	$75.00-$82.00 (R) $112.50-$126.50 (B)
Mississippi	$238.75-$245.75 (R) $267.75-$281.75 (B)	$70.01-$77.01 (R) $112.50-$126.50 (B)
North Carolina	$241.75-$248.75 (R) $264.50-$278.50 (B)	$79.51-$86.51 (R) $111.50-$125.50 (B)
South Carolina	$230.50-$237.50 (R) $253.50-$267.50 (B)	$66.90-$73.90 (R) $111.50-$125.50 (B)
Tennessee	$24.40-$31.40 (R) $85.90-$99.90 (B)	$33.00-$40.00 (R) $111.50-$125.50 (B)

NYNEX

Table A-5 shows pricing for a 2B+D configuration with one B channel alternate voice/circuit-switched data and one B channel circuit-switched data. NYNEX charges business rates for ISDN service even for a line in your home.

Table A-5: NYNEX ISDN tariffs

State	Installation Charges	Monthly Charges
Massachusetts	$158.02	$37.60
Maine	$161.00	$54.20
New Hampshire	$145.00	$46.35
New York	$195.05	$32.23
Rhode Island	$149.61	$48.31
Vermont	$141.00	$81.50

Pacific Bell

Pacific Bell is one of the best telephone companies for ISDN service. They offer a very affordable residential ISDN service, called Home ISDN. Business ISDN comes under their FasTrak ISDN services, which include Centrex and a single business line service. The installation charges for each type of ISDN service listed in Table A-6 includes a one-time charge of $125.00 that is waived if you keep your ISDN service for two years. Pacific Bell charges a penalty for canceling before the two years. In addition, the charges in Table A-6 reflect the optional D channel packet service, which costs $4.00 per month and a one-time $25.00 installation charge.

Table A-6: Pacific Bell ISDN service charges

ISDN Service	Installation Charges	Monthly Charges
Centrex ISDN	$245.00	$35.65
ISDN (single-line business)	$220.75	$28.82
Home ISDN	$184.75	$28.50

Southwestern Bell

Southwestern Bell services Arkansas, Kansas, Missouri, Oklahoma, and Texas. Getting any tariff information from Southwestern Bell was almost impossible. The only information I was able to obtain was from a Southwestern Bell office in Dallas, Texas. The installation charge for an ISDN line in the Dallas area is $556.90. Table A-7 lists the three monthly ISDN service packages for the Dallas area.

Table A-7: Southwestern Bell ISDN service packages

ISDN Service Package	*Flat Rate*	*Notes*
A	$20.90	Includes 15 minutes of local calling. Thereafter, the rate is 25 cents per minute.
B	$30.90	Includes 60 minutes of local calling. Thereafter, the rate is 15 cents per minute.
C	$35.90	Includes 120 minutes of local calling. Thereafter, the rate is 10 cents per minute.

US West

US West offers ISDN single-line service (2B+D) in Arizona, Colorado, Minnesota, Oregon, South Dakota, and the State of Washington, as listed in Table A-8. In the remaining states that US West services — Idaho, Iowa, Montana, Nebraska, North Dakota, Utah, and Wyoming — there is no tariff for ISDN. This means that service may be available on an individual case basis.

Table A-8: ISDN charges for states in which US West offers ISDN service

State	*Installation Charges*	*Monthly Service Charge*	*Notes*
Arizona	$110.00	$69.00	The monthly service charge includes 200 hours of total B-channel usage. Thereafter, the charge is 10 cents per minute.
Colorado	$67.00	$60.00	
Minnesota	$100.00	$69.00	The monthly service charge includes 200 hours of total B-channel usage. Thereafter, the charge is 2 cents per minute
Oregon	$110.00	$69.00	The monthly service charge includes 200 hours of total B-channel usage. Thereafter, the charge is 2 cents per minute.
South Dakota	$80.00	$84.00	
Washington	$85.00	$35.00 $50.00 $63.00	The monthly service charge includes 200 hours of total B-channel usage. Thereafter, the charge is 2 cents per minute.

Appendix B

ISDN Wiring and Powering Guidelines

ISDN wiring is the same as for POTS, but there are certain peculiarities. This appendix is a reprint of several sections of the *ISDN Wiring and Powering Guidelines (Residence and Small Business) Version 2,* published by the National ISDN Users' Forum. It's a handy document for doing your own ISDN wiring instead of paying $50–$75 an hour for a telephone company installer.

Important Safety Information

DO NOT WORK ON YOUR TELEPHONE WIRING AT ALL IF YOU WEAR A PACEMAKER. Telephone lines carry electrical current. To avoid contact with electrical current:

- Never install telephone wiring during a lightning storm.
- Never install telephone jacks in wet locations unless the jack is specifically designated for wet locations.
- Use caution when installing or modifying telephone lines.
- Use a screwdriver and other tools with insulated handles.
- You and those around you should wear safety glasses or goggles.
- Be sure that your inside wire is not connected to the access line while you are working on your telephone wiring. If you cannot do this, take the handset of one of your telephones off the hook. This will keep the phone from ringing and reduce, but not eliminate, the possibility of your contacting electricity.
- Do not place telephone wiring or connections in any conduit, outlet, or junction box containing electrical wiring.

- Installation of inside wire may bring you close to electrical wire, conduit, terminals, and other electrical facilities. EXTREME CAUTION must be used to avoid electrical shock from such facilities. You must avoid contact with all such facilities.
- Telephone wire must be at least six feet from bare power wiring or lightning rods and associated wires, and at least six inches from other wire (antenna wires, doorbell wires, wires from transformers to neon signs), steam or hot water pipes, and heating ducts.
- Before working with existing inside wiring, check all electrical outlets for a square, telephone-dial light transformer and unplug it from the electrical outlet. Failure to unplug all telephone transformers can cause electrical shock.
- Do not place a jack where it would allow a person to use the telephone while in a bathtub, shower, swimming pool, or similar hazardous location.
- Protectors and grounding wire placed by the service provider must not be connected to, removed, or modified by the customer.

An Introduction to Wiring and Powering Considerations for Plain Old Telephone Service

This section addresses the wiring and powering practices used to provide your current (pre-ISDN) phone service also called Plain Old Telephone Service (POTS). For simplicity the words house, home, and residence are used to refer to both a single family residence and a small business. The information is presented at a level of detail suitable for the electrician, wiring installer, or interested homeowner.

The discussions in this section are limited to single line and dual line service installations. It answers questions such as:

- How are the wires run within the house?
- Where does my POTS phone get the electrical power to operate from?
- How can I find the location where the telephone company's wire stops and the internal house wire starts?
- What types of cables am I likely to find in my house?. In addition, a number of terms common to both POTS and ISDN are introduced.

Why are POTS considerations being addressed in a document, whose stated scope is limited to ... wiring and powering practices for ISDN in residential and small business environments? A basic background in POTS concepts is invaluable to understanding the remainder of this document and will aid in the understanding of in-house ISDN requirements.

POTS components

Figure B-1 shows the typical components that are used to distribute POTS throughout the residence. This figure, one of many possible wiring arrangements, is intended to introduce useful terms and wiring components. A more detailed discussion of POTS wiring arrangements is provided later in this appendix. The components and concepts discussed here are.

- The demarcation point (demarc)
- in-house phone cable - also called customer premises wiring and in-house wiring cables
- star, daisy chain, and home-run wiring
- phone outlets (Jacks) and Mating plugs
- distribution devices
- extension bridge
- extension cords and line cords
- FCC Inside Wire Docket 88-57 - also called the 12 Rule
- jack adapters
- POTS terminal equipment

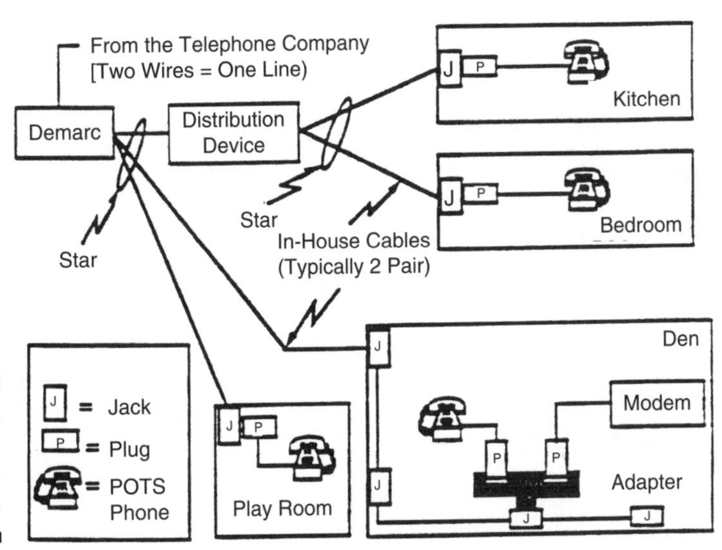

Figure B-1: Typical POTS Components.

Not all of the components described here are readily available in consumer outlets and you may have to order from product catalogs. If you are having a problem locating catalogs, ask someone who works in the electronics or telephone industry for help.

The demarcation point

The single-line service, such as that shown in Figure B-1, is delivered to the house from the telephone company's building (Central Office) on two wires. By law, the telephone company's responsibility ends at the point where these wires are delivered to the residence. The point, where the homeowner's responsibility starts, is called the demarcation point (demarc). In other words, the demarc is the dividing line between the telephone company's wire and the homeowner's in-house wire. The physical device that provides the means to connect the telephone company's wire to the in-house wire is call a Network Interface. In order to reduce complexity, the term Network Interface will not be used in these guidelines. Instead we will refer to both the legal separation point and the physical connection device as the demarc. If you have a two-line service there will be two separate connection points, or demarcs. The demarc may be located inside or outside your house.

While there are differences between demarc devices, there are only two general types. In these guidelines these are called the *Older Type of Demarc* and the *Newer Type of Demarc.* If your in-house wiring has been in place for a long time you may not have a demarc device, in which case your cables will be connected directly to a device called a protector block. The older type of demarc and protector blocks are discussed in section, and the newer type of demarc is discussed in the In-House Telephone Cable section. The demarc(s) and/or protector blocks that are installed in your house may look different than those shown in these guidelines since there are many physical variations of these devices.

The protector block and the older type of demarc

The protector block, shown in Figure B-2 and the upper-left corner of Figures B-3 and B-4, is the telephone line termination device that you are most likely to find in your house. This block may be directly connected to the in-house cables as shown in Figure B-2 or it may be connected to a demarc device such as the service outlet of Figure B-4.

A protector block consists of a plastic block with three threaded posts, each with washers and mating nuts, used as wire connection points and devices that look like two bolt heads (hex or slotted). Not shown is the hardware used to physically attach the block to the house.

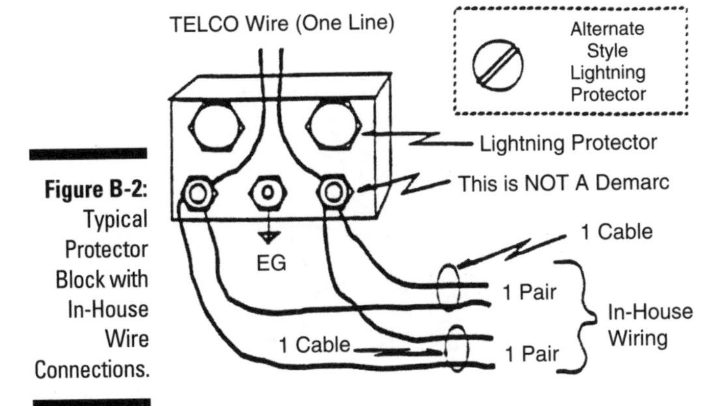

Figure B-2: Typical Protector Block with In-House Wire Connections.

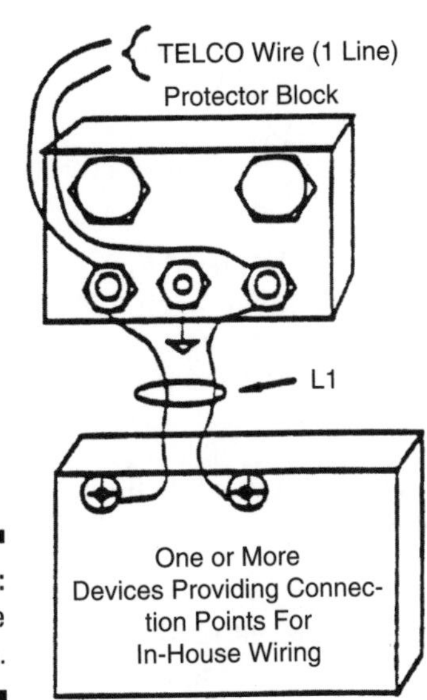

Figure B-3: Older Type Demarc.

Warning: The devices that look like bolts are not bolts, they are *lightning* protectors. ■

It is usually easy to determine which are the in-house wire connection points since the lightning protection devices will not have wires connected. But if you are unsure get professional help.

Warning: Never unscrew the lightning protection devices and never work on your in-house wiring during a thunderstorm. ■

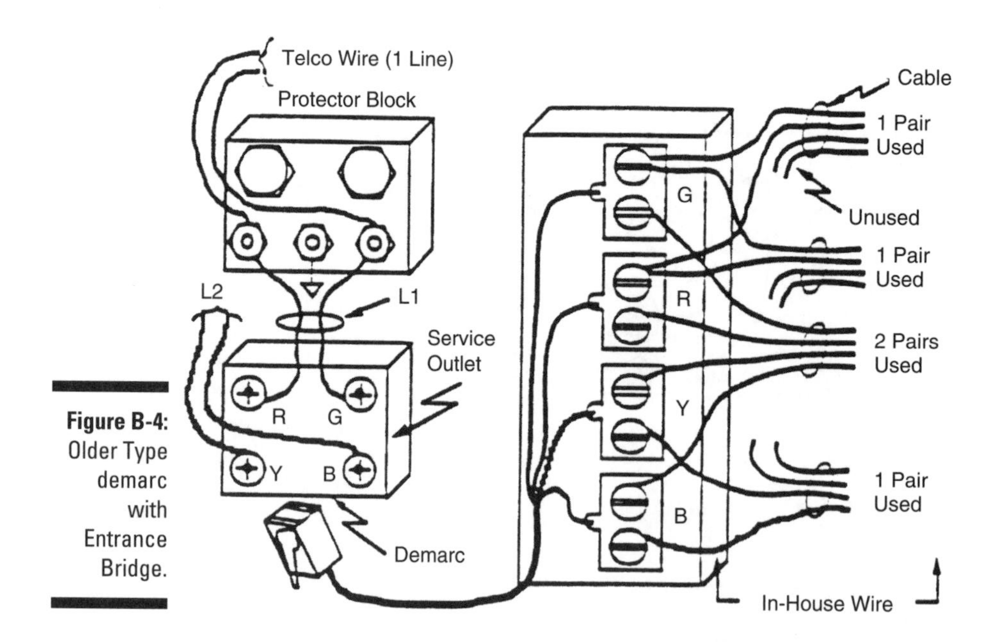

Figure B-4: Older Type demarc with Entrance Bridge.

The center threaded post of the protector block is used for a connection to each ground as part of the block's lightning protection circuit. This wire is connected to ground point, usually a nearby water pipe, by the telephone company telephone company at installation, but it is always a good idea to check that there is a solid connection.

Warning: Never disconnect the earth ground, if you find it disconnected you should contact your local telephone company. ■

The other two threaded posts on the protector block are used to connect your in-house wire with the telephone company's line, either directly as show in Figure B-2 or indirectly as in Figure B-3. There will be a separate protector block for each telephone line coming into your house.

If your in-house cables go directly to the protector block's threaded posts, as shown in Figure B-2, there may be a number of different pairs attached to these posts. In the figure, two pairs (four wires) in two separate cables are used for the in-house wiring and one pair for the telephone company's wire. It is important to understand that in this case there is no legal demarcation point. Before making any changes to your in-house wiring you will have to understand the restrictions associated with the 12-inch Rule.

Warning: If your configuration is like that of Figure B-2. It is important to understand the restrictions related to the 12-inch Rule. ■

In other homes the protector block is wired directly to one or more intermediate devices as shown in Figures B-3 and B-4.

In this case there are usually only two pairs of wires connected to the protector block: one for the wire from the telephone company and one pair (label L1 in Figures B-3 and B-4) connecting the phone line to the device that provides the legal demarcation point. Typically, the device that the L1 pair terminates on provides the type of legal demarc called the older type of demarc in these guidelines. You should understand that while they are called this type of demarc the older type, it is still being installed in many parts of the country for both POTS and ISDN. Figure B-4 provides a specific example of how an older type of demarc might be wired.

Three components are shown in Figure B-4: a protector block, a service outlet and a device call and *entrance bridge*. The service outlet, or similar device, provides the demarcation connection point via a modular jack. The protector block is connected to the service outlet via a single pair of wires labeled L1 in the figure. L1 is connected to the red/green termination points on the service outlet, which correspond to lines three and four of the modular jack and plug. A second phone line, like L2 in the figure, can also be connected to the demarc jack at the yellow/black termination points (pins two and five). The final component shown in Figure B-4 is an entrance bridge, which is a type of distribution device. An entrance bridge has a modular plug and cord on one side and a set of terminals for connecting cables on the other side. In the example of Figure B-4, four in-house cables are shown connected to the entrance bridge:

- The first two cables are connected to L1 via the red/green terminals of the entrance bridge, and each cable has a spare pair.
- The third cable is connected in a typical dual-line fashion to both L1 and L2.
- The last cable is connected to L2 and has a spare pair.

Because the entrance bridge is on the customer's side of the demarc this device is often an option you must request when you order your phone service or install it yourself.

The actual older type of demarc that you find in your house may vary from the configuration shown in Figure B-4. The other possibilities are as follows:

- The protector block and the service outlet combined in one unit. (An external extension bridge may or may not be present.)
- The entrance bridge and the service outlet combined in one unit. (Access to the demarc is preserved by a built-in plug and jack like that shown for the newer type of demarc.)

In addition to the configurations just described, newer installations often utilize a single unit, combining the components shown in Figure B-4. This *new type of demarc* is shown in Figure B-5.

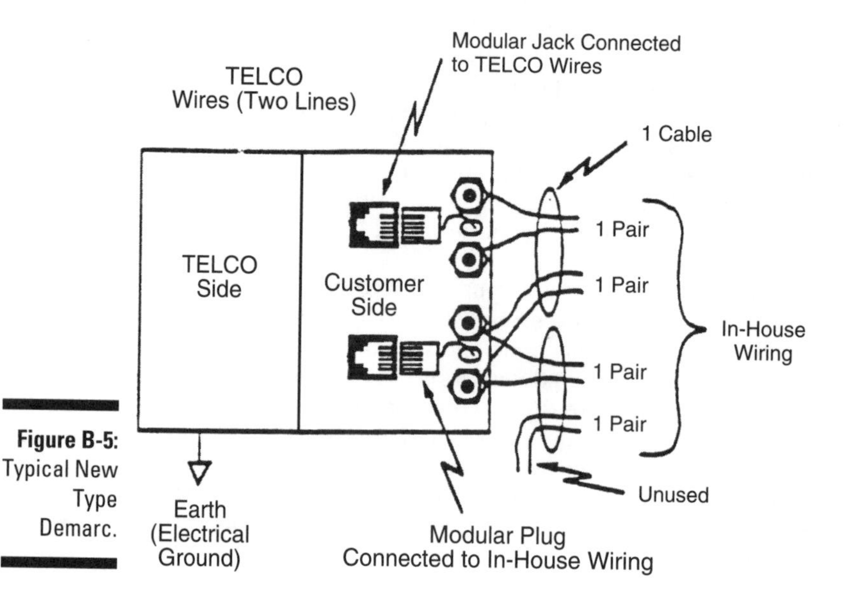

Figure B-5: Typical New Type Demarc.

FCC Docket 88-57

In FCC Docket 88-57, the FCC places an inside wiring restriction, commonly called the 12-inch Rule, on both the telephone company and the customer. If your in-house wiring terminates directly on a protector block, like the wiring shown in Figure B-2, then you need to install a demarc device before proceeding with any wiring modifications. The 12-inch Rule as applied to a single-unit building, such as those buildings that are the subject of these guidelines, can be summarized as follows:

- The demarcation point must be located within 12 inches of the protector block.
- If there is no protector block, the demarcation point must be within 12 inches of where the telephone company's line enters the single-unit premises.
- The customer may not wire directly to a protector block.

Warning: You should not attempt to modify a configuration like the one in Figure B-2 without first knowing how to comply with the 12-inch Rule while providing your own in-house wiring. ■

Newer type of demarc

Figure B-5 shows the newer type of demarc. The unit shown supports two phone lines and is really two demarcs combined in one box. Units are available that support almost any number of phone lines (demarcs). Most likely you will find the one on your house supports one or two demarcs. These units have two sides, one for the telephone company's wire (telephone company side) and one for the in-house wire (customer's side).

The telephone company's wire connection point(s), the lightning protectors, and the earth ground connection point, are not accessible to the user. They are hidden behind a locked door or panel on the telephone company side of the unit. However, a wire does come out of the telephone company side for connection to earth ground. While this wire will have been properly grounded by the telephone company at installation, it is always a good idea to check for a solid connection.

On the customer's side, the connection points are threaded posts with mating nuts, but there is no ground post. In addition, the customer's side has an arrangement for disconnecting the in-house wire from the telephone company wire. This arrangement consists of a modular jack that is connected to the telephone company's wires and a short wire with a modular plug that is connected to the in-house wire connection posts. (Please refer to the Phone Outlets, Jacks, and Mating Plugs section for the definitions of modular jacks and plugs.) The ability to isolate the telephone company side of the demarc from the in-house wire is extremely useful when problems need to be isolated.

In-house telephone cable

Referring back to Figure B-1, we see that wires coming out of demarc are labeled *in-house telephone cables.* These cables carry the wire pairs that are connected to the demarc connection points. Each cable can have one or more pairs of wires. Your house is most probably wired for telephone service with a type of cable commonly called *quad cable*. A quad cable is shown in Figure B-6. It has two pairs of wires enclosed in a outer plastic jacket or sheath.

The type of telephone cable used in your house is dependent upon when your house was built and what modifications have been made since then. Some, but not all, of the possibilities are listed as follows:

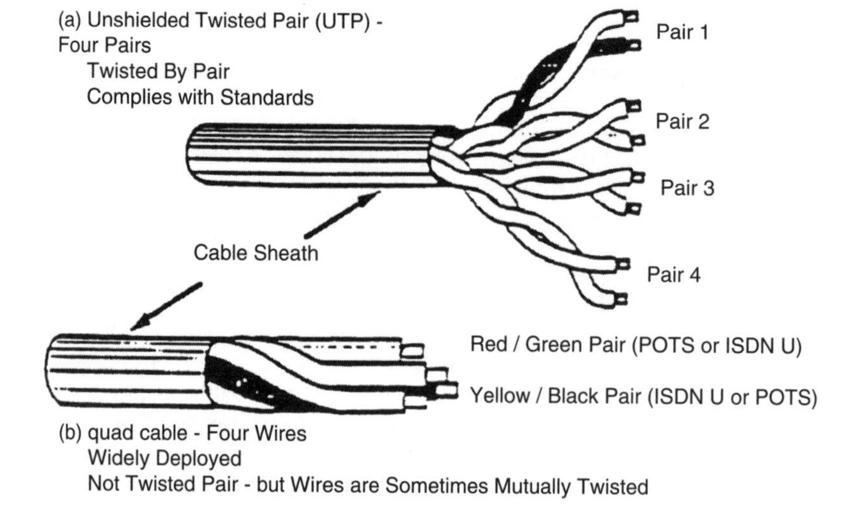

Figure B-6: Quad and unshielded twisted pair cables.

- Prior to the early 1950s your telephone company installed wire without an outer jacket. This type of unjacketed cable was available in two wire (one pair), three wire, and four wire (two pair) versions. In order to hold the wires together in a cable the wires were twisted together. All of the wires had the same color insulation, usually white, and color coding was done with a fabric inside each wire's insulation. The color coding was the same as that used in quad cable: Red/Green for the first pair and Yellow/Black for the second.
- In the early 1950s, phones with lighted dials were introduced and the Telephone Companies started to install quad cable. This wire, shown in Figure B-6, is still being installed by some Telephone Companies and independent installers.
- In the early 1980s customers were given the right to select an independent contract or for in-house telephone wiring. The large majority of these independent contractors continued to use quad cable, as did the Telephone Companies. However, variations in the characteristics of quad cable have increased in recent years and some of this cable is inferior to that used by the Telephone Companies. There are a growing number of new homes being wired with this lower-quality quad cable as well as other types of inexpensive cable that are not suitable for two-line POTS installations of ISDN. As a result of the lack of availability of clear guidelines for cable selection, many POTS customers are experiencing noise problems. Another common problem that is directly related to the installation of low-quality cable is interference between lines (crosstalk) in two-line POTS installations. In-house cable that is not suitable for POTS will not provide a quality ISDN connection.

- There are also many instances of homeowners running their own wire to add or move phones. Often the wired used in these cases has been determined by what has been readily available. It is safe to say that almost every conceivable type of wire is in use someplace.
- In the early 1990s national standards (EIA/TIA wiring standards) were adopted for communications wiring practices including POTS and ISDN. A discussion of these standards is beyond the scope of this document but it is important to know that they recommend that all new homes be wired with Unshielded Twisted Pair (UTP) cable.
- A twisted pair consists of two wires (a pair) twisted together. An unshielded twisted pair cable contains two or more twisted pairs enclosed in a jacket or sheath. An unshielded pair cable with four twisted pairs (eight wires) is shown in Figure B-6. The designation unshielded refers to the fact that the sheath does not include an electrical shield.

  Warning: Cables with shields are often used to eliminate unwanted electrical interference, but they should never be used for POTS or ISDN. ■
- The EIA/TIA Standards classify types of UTP cables as Category 3, Category 4, and Category 5 in accordance with their electrical properties. The minimum requirement, specified by the standards, for new residential wiring is Category 3 UTP, but there has been little or no compliance. quad cable is not EIA/TIA standards complaint.

Star, daisy chain, and home-run wiring

In Figure B-1 three cables are shown leaving the demarc in a configuration known as a *star configuration*. The configuration is called a star because a common connection point supports wiring to more than one location. In the following discussion, it is assumed that the house is wired with quad cable that has two pairs. Since this is a single-line service, only one pair from each cable is connected to the common point at the demarc. At each service outlet (jack), however, all four wires should be connected.

The cable running to the play room goes directly to a single jack called a home-run. Star wiring is also often referred to as home-run wiring.

The cable going to the den is connected to four outlets in a configuration know as a *daisy chain.* In a daisy chain, the cable in terminated at the first outlet (jack) where a second length of cable is also connected in order to continue the run to the next outlet. In this case, the outlet is serving a dual purpose as a phone jack and a distribution device. This wiring continues until the last outlet in the daisy chain. In Figure B-1, the daisy chain is shown as confined to a single room. This is not always the case and your house may be wired as one large daisy chain. Additional information on POTS wiring configurations is included later.

Distribution devices

In Figure B-1, the cables running to the kitchen and the bedroom are connected to a *distribution device* where they are joined to a single cable from the demarc. While there are many different type of distribution devices, they are all used to connect a single-cable input to multiple-cable outputs.

An important characteristic of a distribution device is the ability to provide reliable connections, using mechanical connection schemes (no solder), that resist corrosion. Another function of a Distribution Device is to provide an easy way to move and/or add cables. It is poor practice to connect cables by twisting the wires together and/or using solder.

The *junction box* shown in Figure B-7 is similar to the surface-mount outlet box with screw-down terminals, however, it has no jack. Typically, each connection point consists of a screw with two washers. Because good practice dictates that a washer be used to separate each wire, this device should only be used for one two-pair cable input and two two-pair cable outputs. Remember, both pairs in the cable should always be connected through to the final destination, even for single-line service.

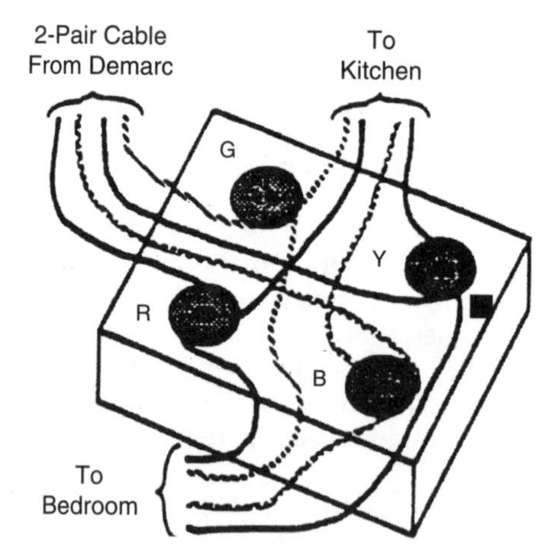

Figure B-7: Junction Box.

The entrance bridge of Figure B-8 is a device that is often used for connecting to an older type of demarc. It should also be mentioned that the newer type of demarc includes a built-in entrance bridge and thus also provides the functionality of a distribution device. The entrance bridge shown in this Figure provides reliable connections for twice the number of wires that can be connected to a junction box. Other styles have provisions for additional connections. Both the junction box and the entrance bridge, including the built-in entrance bridge that is part of the new type demarc, limit the number of cable connections and lack the flexibility needed in some applications. In such cases, the entrance bridge or junction box can be connected to one or more additional distribution devices of a different or like kind.

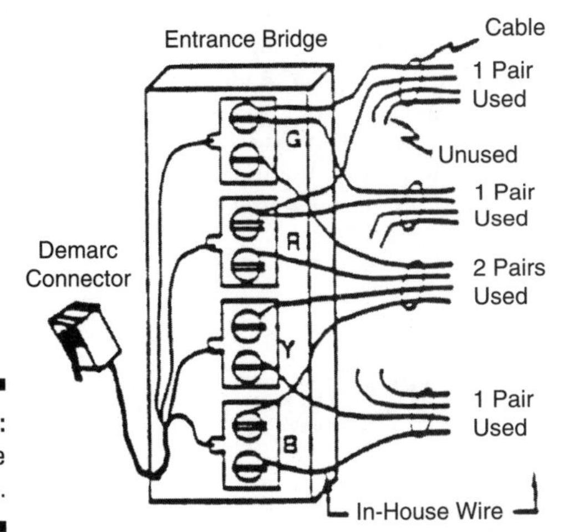

Figure B-8: Entrance Bridge.

An example of the use of multiple-distribution devices; is shown in Figure B-1. In this configuration, a distribution device, built into the demarc, provides connections for three cables in a star configuration where one leg of the star is connected to a second distribution device. Cascading co-located distribution devices is a technique that can be used to provide low-cost flexibility. However, caution is advised when installing cascaded distribution devices because increasing the number of connection points reduces the overall reliability of the in-house wiring.

Note: Cascaded distribution devices are not recommended for ISDN installations. ■

The bridging block of Figure B-9 has a number of threaded post-pairs embedded in a plastic block. These posts, with their associated washers and mating nuts, are like those found on some newer type demarcs. The block shown is wired for use in the configuration of Figure B-1. Each pair in the cable from the demarc terminates on one pair of posts which in turn is connected to other post pairs, by jumpers. All wires are separated by washers. In the figure, the service is continued to the kitchen and bedroom using separate post-pairs, but multiple pairs can share a single post-pair — provided that the wires are separated by washers. It is good practice to connect all pairs in a cable through to the final destination (jack).

A distribution device commonly known as a *66-block* is usually found only in large building installations. Occasionally they are used in residential and/or small business installations. Because they provide more flexibility than bridging blocks, entrance bridges, and junction boxes, their use in residential and small business installations is likely to increase. A variation on the 66-block provides modular jack outputs. The use of these devices for wire distribution is beyond the scope of this section. As they are seldom found in residential installations, they require a special tool, and practice is needed to make good connections. This discussion of 66-blocks has been included for completeness and to introduce a connection scheme that is similar to a connection method used in some newer phone outlets (see section on Phone Outlets, Jacks, and Mating Plugs). If you run across a 66-block in your house it is best to seek the advice of a skilled installer.

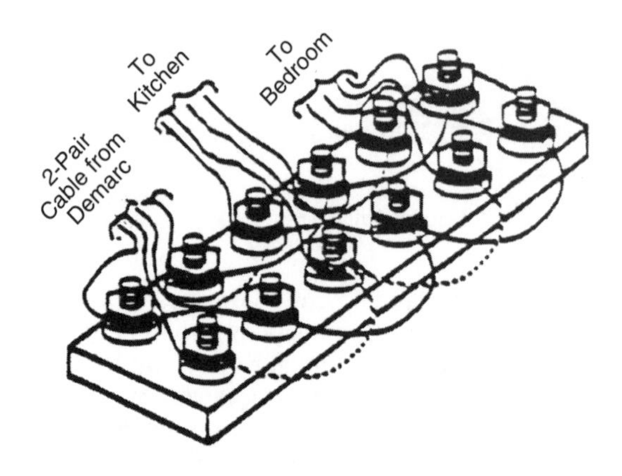

Figure: B-9 Bridging Block

The method used to make wire connections on a 66-block is called *insulation displacement.* A typical insulation displacement connection has a V-shaped groove. An excellent connection is formed by forcing the wire, insulation intact, into the groove with the special tool. The groove is designed to remove the insulation, without breaking the wire, while compressing the copper conductor. The result is a metal-to-metal connection, called a gas-light connection between the wire and the displacement connector. This connection is safe from the corrosive effects of the atmosphere.

A *patch panel* is a distribution device used primarily in large installations, where rearranging the wiring is an important attribute. Patch panels are not presently being used in small business and residential installations but, because of the flexibility they provide, they are worth consideration for new installations and may become common in the future. This discussion of patch panels has been included because their flexibility is attractive for new installations.

A patch panel is just one of many configurations and it can be used to provide connections for all home wiring applications. In this case two cables are shown as inputs, one for ISDN and one for POTS. The input cables are wired to the top row of modular jacks at the back of the panel. The POTS cable is connected to jacks one, two, and three the ISDN input cable is connected to jacks four, five, and 6. Cables to service outlets are wired to the bottom row of jacks, one cable for each position. The desired service is directed to any specific outlet by connecting patch cords between the upper and lower rows. Patch panels that comply with the EIA/TIA wiring standards are expensive in comparison to the cost of other distribution devices, but the flexibility they provide may justify their use in the home of the future.

Though we have described some common distribution devices, we have not tried to provide an exhaustive list and you may find other types installed in your house. It is not uncommon, for example, to find a phone outlet used as a distribution device with or without a phone attached.

Phone outlets, jacks and mating plugs

If your phone installation is old, you may find that there is no phone outlet. In the older installation a wall phone was hard-wired directly to the demarc and a desk phone was wired to a junction box.

Phone outlets were introduced by the telephone companies to allow the phone to be moved between rooms, as well as to provide a way to quickly disconnect the phone for service. The early type of phone outlet has four cylindrical female (jack) contacts located on a surface mount block (wall or baseboard mounting). The mating plug four pins (male connectors) arranged to match the jack's pattern. An adapter plug that provides conversion to the newer type of plug called a modular plug.

The type of outlet that you are most likely to be familiar with is shown in Figure B-10. It is a surface-mount unit with a jack, called a modular jack, on the side. The mating modular plug is also shown.

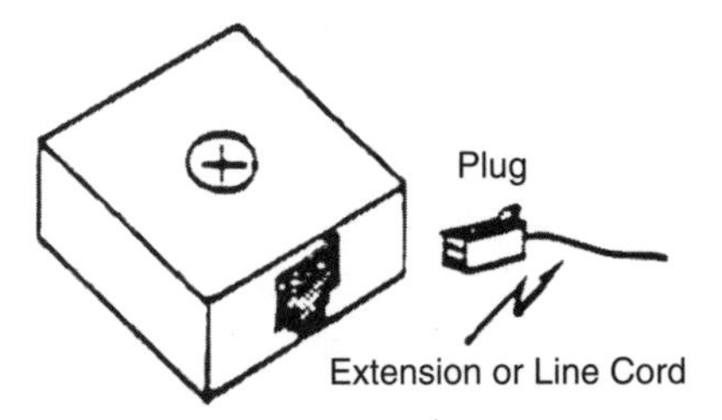

Figure B-10: Surface, mount, outlet modular jack, and plug.

The newest type of service outlet is the flush mount type, shown in Figure B-11. This type of service outlet is mounted in the same way that electrical wall outlets are mounted. The service outlet shown in the figure has two modular jacks, but other variations are available. Flush mount service outlets are commonly used in new construction.

While the terms RJ11 Jack and RJ11 plug are commonly used to describe these modular components, these designations are not always correct. The terms six-position modular jack and six-position modular plug are used in these guidelines.

Note that the term six-position modular jack (plug) does not always imply that all six metallic conductors are present. The variations use the same modular jack (plug) housings but all conductors (pins) are not always present. The 6 x 4 jack is the variation that you are most likely to find in your house. The notation 6 x 4 indicates that the jack is six positions wide with only the four center positions installed. The line cord connecting the phone to the jack will most likely have a 6 x 4 or 6 x 2 plug.

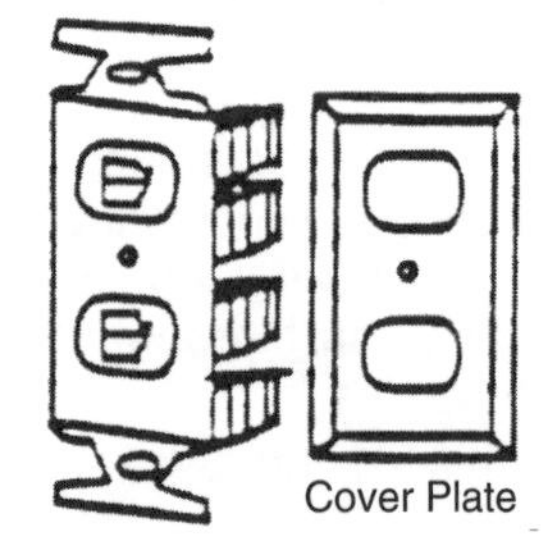

Figure B-11: Flush, mount outlet with modular jacks.

As you will see later, ISDN installations sometimes use the six-Position Modular Jack but more typically use a eight-position modular jack, which only has one configuration. The eight- and six-position modular plugs will mate with the eight-position modular Jack. Some caution will be required to ensure proper use in an ISDN installation.

Note: For all renovations and new installations, jacks and plugs that conform to the EIA/TIA standard's Category 3 or better are recommended. ■

Making connections to phone outlets

There are two types of wire connection methods used to terminate wires on the jacks used in phone-service outlets. If you have an *early outlet plug*, then there will be screw-type termination points. These will have markings labeled

R, G, Y and B to indicate the wire color code. The Red (R) and Green (G) pair is always the primary (first) phone line and the Yellow (Y) Black (B) pair the second phone line. Homes with this type of outlet usually have only one phone line present at the jack even if there are two lines coming into the residence. If you are installing ISDN service at this jack, you will have to replace it with a modular jack.

The service outlets shown in Figures B-10 and B-11 may have screw-down terminals or punch-down terminals. The screw-down terminals are similar to those described for an older outlet, but the jack are commonly used for both two-line and one-line service. The punch-down terminals are the insulation displacement type for the 66-block. However, in this case, the tool is built into the jack housing and no special skill is required. Just follow the directions.

Jack wiring, or which wire goes where?

If the outlet supports one single-line phone, then the Red and Green pair is connected to jack pins three and four. The Yellow/Black pair will also usually be wired to the jack. If the jack supports a single-line phone for a second line then the Yellow/Black pair is used. Note that two separate single-line phones can also be connected to this jack if an adapter is used. See section on Jack Adapters for more information.

Extension cords and line cords

These two types of cords are similar: they are both used to connect the phone to the service outlet. A line cord is a flexible cord with plugs on both ends. It is used to directly connect from the service outlet jack to the jack on the phone. Extension cords have a plug on one end and a jack on the other end; they are used to extend the reach (distance) of a line cord. There is almost no limit on the length of either a line cord or an extension cord used for POTS, but this is not true in ISDN implementations.

Another important characteristic of these cords is that, for historical reasons, they reverse the pins on each pair from one end to the other. A two-pair reversed cord is sometimes called a crossed cord. Straight-through cords are required for certain types of ISDN connections.

Warning: Many cords that look alike have wires that are connected to different modular pins than those in the above examples. Always check the suitability of cords for the intended ISDN or POTS use. ■

Jack adapters

Adapters are used to convert the wiring at a service outlet jack into another wiring configuration. They are also used to provide fan-out connections (one-to-many) and convert from six (eight) position modular to eight (six) position modular configurations.

An adapter that converts one jack into four jacks usually provides a direct pin mapping (pin 1 to pin 1, etc.) and is used to provide parallel connections allowing a single jack to fanout to multiple devices.

Warning: It is not good practice to run cords long distances. They are not designed for this sort of use. This is especially true for ISDN service. You should always use cables for wiring between the demarc and the service outlets. ■

In Figure B-1 an adapter is shown supporting a phone and a modem in the den. Often, devices like modems have built-in adapters, eliminating the need for an external adapter.

An adapter that converts a six-position modular jack to an eight-position modular jack is often useful for converting existing POTS jacks to ISDN jacks.

Another type of adapter is used to convert a service outlet wired for dual-line service into two jacks that are wired for single-line service.

The adapters described here are only a sample of what is available. Most are only available from parts distributors' catalogs. Warning: many adapters look alike but provide different pin translation (in-to-out).

POTS terminal equipment - also called customer premises equipment

In the above descriptions we have used the term telephone (phone) in a very loose fashion to mean any device that is connected to a POTS line. However, there are a number of other devices that are used for voice and data communications over a POTS line. These are facsimile (fax) machines, answering machines, alarm systems, calling number identification boxes and modems. Two general terms are commonly used to describe phones and these other communications devices: terminal equipment (terminal equipment) and customer premises equipment (CPE). In these guidelines you will find both of these terms used frequently.

Powering terminal equipment

Your POTS telephone gets the power required to operate from the telephone company. This power is delivered to your house over the same pair of wires that carry your voice. However, if you are a user of any other POTS terminal equipment, for example a cordless phone, it probably needs to be plugged into an AC outlet. Even some new phones, those with fancy features, will not operate properly unless a small power supply is plugged into the house's AC power.

A basic POTS phone is a very simple device. It requires no power when not in use (on-hook) and very little when it is operating (off-hook). In contrast, other terminal equipment requires considerable power to operate even in stand-by mode (on-hook). The power system that feeds your phone over the telephone line was designed for the POTS phone and is not capable of driving many types of modern terminal equipment.

POTS wiring configurations

Two basic types of wiring configurations were introduced in an earlier section: the star and the daisy chain. This section introduces a third configuration and expands on the information previously provided.

The *ring configuration* is an extension of the daisy chain configuration, with the wiring from the last outlet returned to the distribution device. This wiring scheme provides redundant connections so that the terminal equipment at all the service outlets will survive a single cable break. While this is a good configuration for POTS wiring, it cannot be used for some ISDN or other modern communication services.

The new national wiring standard (EIA/TIA-57.0) recommends the use of the star configuration in all new residential construction and renovations to all existing residential wiring. Because these standards are new, the level of compliance is still low.

What you will find in your home may not look like any of the configurations presented in this section. Instead, the wiring most likely will be a composite of other configurations shown. Figure B-1 is an example of composite configurations.

When you investigate how your house is wired, you may find some practices that we have not described in these guidelines, as well as some of the following conditions. You may find some hidden unused jacks under counter tops, in closets, and other unlikely places. There may be abandoned wire where the jack has been removed but the wiring is still connected to the demarc. Another type of abandoned wire is where the jack is still in place but the demarc side of the cable is disconnected. Abandoned wiring and wiring to unused jacks are called *bridged* taps. A bridged tap is a wire run that is not terminated by the presence of terminal equipment or other devices.

The homeowner may create a distribution device by simply twisting wires together mechanically and/or soldering them together. This type of connection is bad practice in a POTS environment and must be eliminated in an ISDN installation.

Other things that you may find in your home wiring include:

- only one pair continued from outlet to outlet in a daisy chain.
- the wrong wires used as a pair (pairs must be Red/Green and Yellow/Black).
- wires run parallel to the AC power cables.
- types of wire not described in this section. For example, an extension-phone outlet wired with electrical cord, audio, speaker wire, thermostat cable, or other odd ball cable.
- flat cable used to connect a wire run under a rug.

An Introduction to ISDN Wiring Considerations

This section provides information about ISDN wiring considerations for a residence or small business. It assumes that the reader may be purchasing and/or installing ISDN wiring and equipment. Understanding how the ISDN components work together prior to starting the purchasing and installing process should help the process go smoothly. The objective is to introduce and provide simple explanations for a variety of topics related to ISDN wiring issues, such as:

- ISDN connectors
- ISDN outlets
- ISDN cords
- ISDN cable
- demarcation point (network interface)
- U-interface
- Network Termination 1 (NT1)
- terminating resistors
- NT1 timing
- S/T-interface
- terminating equipment
- extension phones and the R-interface

ISDN wiring requirements

Wiring considerations for ISDN are different from those for POTS. Wiring for POTS is usually simple and straightforward. For example, normally it is possible to add an extension telephone in a home simply by adding another jack and plugging in the telephone. The wiring for the new jack may have been spliced into existing wiring, or perhaps extended from an existing jack in order to accommodate the new jack. Whatever the situation, if one follows the color code and exercises reasonable caution, it is likely that things will work fine. Similarly, it doesn't make much difference if there are unused jacks around the house. However, there are some new factors to consider with ISDN. The ISDN electrical signals that must be transported in the home are different from those for POTS. The wiring used for POTS may or may not be suitable for ISDN. The characteristics of the cable, the manner in which it is installed, and the placement of components along the cable can affect the quality of the service. One also needs to understand whether, in certain cases, existing wiring might be usable for ISDN or if new wiring required.

In addition to how the equipment functions and how the house is wired, there are also differences between POTS and ISDN with respect to how the equipment (a telephone) is powered.

ISDN Components

connectors

Before we attempt to discuss ISDN telephones and other similar components, it is important to spend some time on the most basic items involved in providing proper wiring for ISDN. One of these basic items is the so-called modular connector (modular plugs and modular jacks) that have been used in recent years by the telephone industry. There are two that are of particular interest — the modular six-position jack/plug and the modular eight-position jack/plug. The modular six-position jack/plug is often referred to as an RJ-11 jack/plug, and the modular eight-position jack/plug is often referred to as an RJ-45 jack/plug. While the RJ-11 and RJ-45 designations are commonly used, there are situations when they might cause some confusion, so we will use the more generic six-position and eight-position terminology here. The connector normally used for ISDN service/equipment is the modular eight-position jack/plug. Although not typically used for POTS in a residence, modular eight-position connectors are often used for data services. Unlike the modular six-position plug/jack where some of the pins are often not present, the eight-position plug/jack typically has all eight positions equipped with pins.

We have briefly mentioned the standard connectors (modular six- and eight-position plugs and jacks) and several of the wire types that one might have to deal with in wiring for ISDN. Now we would like to introduce some examples of the additional nomenclature that arises when these plugs/jacks are wired for certain applications - you may or may not encounter such nomenclature. A modular eight-position plug/jack with only the center two pins wired is referred to as a T568A connector. However, a modular eight-position plug/jack might also be a T568B connector — similar to a T568A except with pairs two and three swapped. The applicable color coding for wiring an ISDN connector using the recommended four-pair twisted wire is as shown in Table B-1.

It is worth noting that, despite the difference in width, a modular six-position plug can be inserted into a modular eight-position jack. The construction of the six- and eight-position connectors is such that the pins in the six-position plug will align with the pins in the six center positions of the eight-position jack. For example, if the center four pins are present in a six-position plug they will align properly with the center four pins in an eight-position jack. While the use of a six-position plug in an eight-position jack is not recommended or encouraged, it can be used without fear of mismatching between the plug and jack pins.

Table B-1: Premises wiring color code

Conductors	*Twisted Pair Color Coding*	*EIA/TIA Standards 8-Position Pin Number T568A/T568B*	*EIA/TIA Standards 6-position Pin Number*
Pair 1	White/Blue*	5/5	4
	Blue**	4/4	3
Pair 2	White/Orange*	3/1	2
	Orange**	6/2	5
Pair 3	White/Green*	1/3	1
	Green**	2/6	6
Pair 4	White/Brown*	7/7	
	Brown**	8/8	

* The wire insulation is white with a colored marking (typically a stripe) added for identification.

** A white marking (typically a stripe) is optional.

outlets

The recommended jack/wall outlet for ISDN service is the modular eight-position jack referenced earlier. If that type of jack is used wherever there should not be any connector compatibility problems since that jack can receive either an eight-position or a six-position modular plug. It may be possible to use existing six-position jacks/wall outlets, but there is a high likelihood of encountering compatibility problems (your ISDN components will likely have eight-position plugs).

It is possible to buy or make adapters to deal with almost any incompatibility problem, but it seems a more realistic approach to simply use the proper jacks to start with.

cords

There are two types of cords commonly used with terminal equipment: one is a line cord and the other is an extension cord. Although cords are relatively simple, there is a difference between POTS and ISDN cords. A pair-reversal arrangement is usually built into a POTS cord, but such a reversal is not acceptable for ISDN cords — they must be wired on a straight through basis. A pair reversal is not a problem on the U-interface portion of the layout — if, for example, a reversal type cord was used between a U-interface wall jack and an NT1. The reversal problem arises in connection with the S/T-interface if multiple terminals are connected to it. The polarity of any individual terminal is not critical, but it is critical that all terminals on the same S/T-interface be wired using the same polarity. The simplest way to avoid cord-related problems is to avoid the use of reversal-type cords for ISDN. ISDN technical specifications call for limiting the length of ISDN line cords to 33 feet or less.

cable

In addition to cords and connectors, another basic ingredient for providing ISDN service in a home or business is the cable itself. Until the early 1980s, the telephone company was responsible for telephone wiring in homes and businesses. Subsequently, customer premises wiring has been the responsibility of the home/building owner. Even when the telephone company was responsible for wiring, multiple cable types were used everywhere for residence wiring. Since that time there has not been any standard type of cable used in wiring a home.

A wide variety of telephone cabling already exists in homes. Even quad cable, that has been widely used by the telephone industry, has no single standard.

Thus, if your ISDN installation plans include the possibility of using existing wiring, caution, is recommended. There is no simple way to evaluate existing cabling — it may or may not be satisfactory. Probably the best way to determine whether existing wiring is satisfactory for ISDN use is to try it. If there is any doubt, use new wiring as specified in the next paragraph.

If new cabling is to be installed in either an existing home or a new home, it is strongly recommended that the standards set forth by the Electronic Industries Association (EIA) and the Telecommunications Industry Association (TIA) be followed. The EIA/TIA recommended minimum wiring is eight-conduction (four pairs) unshielded twisted pair, category three or higher, 24-gauge. Category 4 or 5 cable is satisfactory for ISDN service, although it exceeds the requirements necessary for ISDN and is somewhat more expensive than category three cable.

The demarcation point

ISDN is delivered to a home or business in a manner that appears identical to existing POTS — a pair of wires that come to the house and connect to a junction box or terminal block of some sort. Typically, that occurs just before or after the wire enters the building. Whether it be a junction box on the outside or terminal block on the inside, the point at which the wire from the telephone company is connected to the wire that runs through the house is usually called the demarcation point. This point is also sometimes known as the network interface (NI) since it is where the responsibility of the network service provider (typically the telephone company) meets or interfaces with the responsibility of the home or building owner. In other words, the ISDN demarc point is the same as the POTS demarc point. The demarc can take many forms — several examples are shown in Figures B-2 through B-5.

ISDN equipment and interfaces

With normal telephone service (POTS), nothing really changes between the demarc point and the terminal equipment (the telephone set) — the same electrical signals that arrive at the demarc from the telephone company central office continue on to the telephone set over two wires. However, with ISDN, an additional component is required between the demarc point and the ISDN terminal equipment. That additional component is called a *network termination* 1 (NT1).

U-interface

As indicated previously, there is much jargon associated with ISDN. One common term is *interface.* One of the interfaces frequently referred to is the U-interface. The U-interface is simply any point along the pair of wires from the telephone office after they have passed through the demarc point and before they get to the NT1. Thus, the term U-interface does not refer to a specific connection point like the demarc, but rather represents certain electrical characteristics, defined by the ISDN technical standards, that occur on a two-wire transmission path.

NT1 and S/T-interface

One side of the NT1 connects to the U-interface. The NT1 converts the signal arriving on the two-wire U-interface to a four-wire electrical signal known as the S/T-interface. The NT1 may be almost adjacent to or a considerable distance from the demarc. There are several significant points worth noting at the outset of our discussion:

- The NT1 may be a stand-alone piece of equipment or it may be built into an item of terminal equipment such as an ISDN telephone. An ISDN terminal with a built-in NT1 is sometimes called a U-interface terminal.
- There can be only one NT1 connected to a U-interface. Thus, if multiple terminals are to be served by one ISDN line, they must be served by a single NT1. Multiple U-interface terminals (with built-in NT1s) are not acceptable.

Note: There are ISDN terminals that have a built-in NT1 with a jack that allows adding another terminal that does not have a built-in NT1. Thus, a stand-alone NT1 is not the only possible option when multiple terminals are involved. ■

For purposes of this discussion, the basic function of the NT1 is to convert the ISDN electrical signals that travel from the telephone office to the NT1 over the equivalent of one pair of wires (two signal leads), into an electrical configuration that uses two pairs of wires (four signal leads), as required by the ISDN customer terminal equipment. The wiring that runs between the NT1 and the terminal equipment is called the S/T-interface. Just as with the U-interface, the term S/T-interface does not refer to a specific connection point, but rather represents specific electrical characteristics. Also, recall that in some applications the NT1 may be included in the terminal equipment so there is no exposed S/T-interface. Unlike with the U-interface, multiple terminal equipment devices can be connected to the S/T-interface.

The ISDN electrical signal that passes from the NT1 to the terminal equipment requires two pairs of wires (i.e., the S/T-interface uses four-wire transmission). However, the S/T-interface is defined as an eight-wire interface — four wires for signal transmission on four wires for optional power arrangements. Depending upon the powering arrangements used, the S/T-interface for a given installation may require only four or six of the eight wires that are specified.

Terminating resistors

In order to ensure the proper electrical characteristics of the signals passing over the S/T-interface, ISDN technical specifications require the use of terminating resistors. Improper placement of the terminating resistors will result in unsatisfactory operation of the equipment. A detailed discussion of terminating resistors is beyond the scope of this section but is covered in later sections. The manufacturer's literature should also provide information on this subject. Figure B-12 provides some examples of the various configurations you may encounter.

NT1 timing

Many NT1s allow the selection of one of two timing modes — fixed or adaptive. A discussion of NT1 timing is beyond the scope of this document, but the manufacturer's literature should provide helpful information when needed. In general, for the types of configurations covered in this document the use of adaptive timing will always be satisfactory.

Terminal equipment

From the customer's perspective, the terminal equipment is where the action is. Terminal equipment is the communication device, or devices, that the customer actually uses. In the POTS case, the terminal equipment is usually a telephone, but it could be a facsimile (fax) machine, a modem/computer, or whatever the customer had connected to the line to handle the desired communication. With ISDN, the concept is the same but the equipment is a bit different - it could be an ISDN telephone, a personal computer with a special plug-in board to provide an ISDN interface, an ISDN fax machine, etc. There are many possible terminal equipment arrangements.

There is a somewhat special situation that is worthy of a brief mention here. The ISDN technical literature defines a device or function known as a terminal adapter. A terminal adapter converts the ISDN signal into a form usable by non-ISDN terminal equipment. Terminal adapters are intended for both data and voice applications. A terminal adapter can be a standalone item or it can be built into a piece of non-ISDN terminal equipment. The signal between the terminal adapter and the non-ISDN terminal equipment is defined in ISDN technical specifications as the R interface.

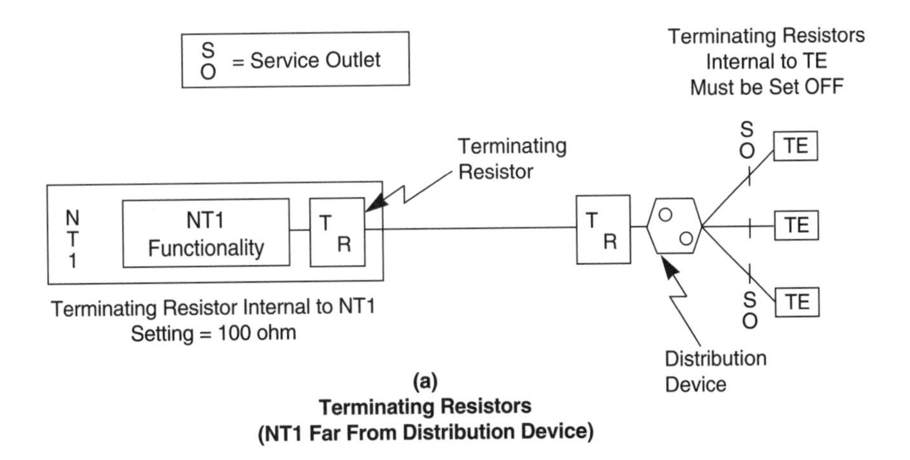

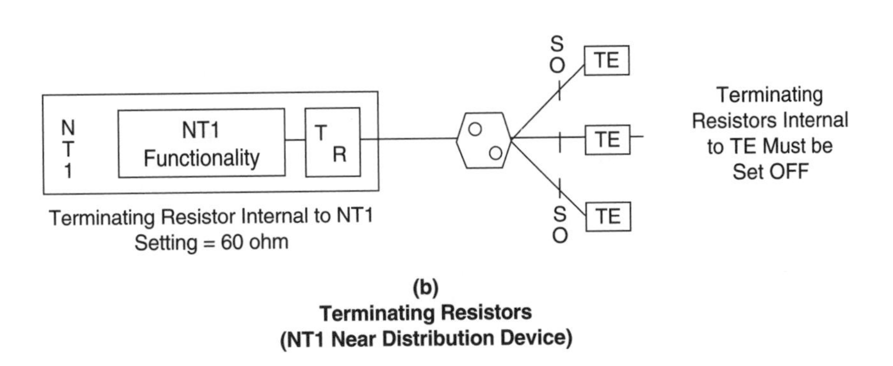

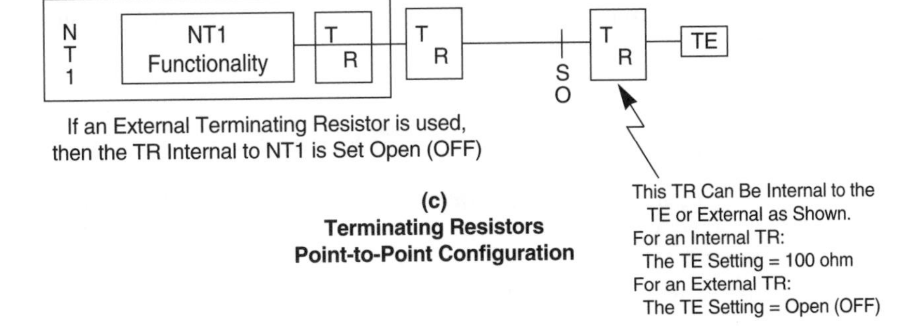

Figure B-12: Examples of terminating resistor usage.

Extension phones and the R-interface

One of the differences between POTS and ISDN is that ISDN does not include the same concept of extension phones as we know in POTS. We cannot just lift the handset of any ISDN phone in the house and have a conversation. A way of providing POTS-like extension capability in an ISDN installation is by using the R-interface mentioned in the preceding paragraph. Unfortunately, the R-interface is not as well defined as it might be. Characteristics such as the number of POTS extension phones supported by an R interface, the number of ringer equivalences, and the allowable cabling distances are left somewhat to the vendor's discretion. Thus, if you need to make use of the R-interface and have specific requirements, it would be well to check them out with the equipment vendor ahead of time.

Working Out a Wiring Plan

For POTS, one can normally add a new extension phone by simply tapping into the existing telephone line somewhere, running the new cable to wherever the new phone should be, and installing a jack to plug the phone into. Other than following the appropriate color code to make the proper connections, the process is quite straightforward. Another extension could be added by again tapping into the wiring almost anywhere and repeating the process. Another type of arrangement often used for wiring new homes is to run the wire so that it passes sequentially through each jack location. This is often referred to as a *series* or daisy chain configuration.

While the series or daisy chain wiring arrangements are acceptable for POTS, they are not consistent with the ISDN technical standards. In some cases, a series-type arrangement will work while in others it will not. The wiring arrangement recommended by the EIA/TIA wiring standards is a Star or home-run configuration, where the wiring to any individual piece of terminal equipment comes from a central or home point and goes only to that terminal equipment location.

New or existing wiring?

When considering the various wiring issues, one of the main questions is whether to use existing wiring, if your situation permits, or install new wiring. You will need to weigh the value of time and materials for new wiring against the value of time to figure out the existing wiring, and plays any risk that the existing wiring might not do the job. If the decision is close, or if there is any significant uncertainty about the suitability of the existing wiring, then new wiring is recommended.

If you will be installing new wiring from the demarc point, it is reasonably simple and straightforward to figure out what you need to do and then do it. New wiring will ensure that there are no unwanted connections along the path of the wiring and, assuming you use cabling of the recommended quality, that the cabling will perform as desired. Thus, the wiring already in place in the house is not a factor. In that case you can ignore the rest of this section that talks about how to deal with the existing wiring. This Appendix provides useful information regarding new wiring later on.

Existing wiring

If you are considering using existing wiring for ISDN, there are some additional factors to be considered. Things like bridged taps or unused jacks can cause trouble in some cases. If you are working with your own home you may have a good understanding of the existing wiring, but if you are working with someone else's home things are likely to be more of a mystery. If you are lucky enough to be working in a house where the wiring is exposed (a ranch house with a full basement where the telephone wires is visible in the basement, much of the problem is solved.) Unfortunately, if most of your wiring is hidden from view the issue gets more complex.

First find out what wiring exists and where it goes. Obviously, if you will be wiring a house that you are not familiar with, one of the first things to determine is how many POTS lines enter the house. Each separate line would have a different telephone number. It is important to keep track of the different lines. If quad wire is used to feed any given jack you should determine whether that particular wiring/jack serves more than one line. Even if you know that a specific segment of quad wire carries only one telephone line, there is always a possibility that the additional wires could carry power for an older-style lighted dial phone, service a burglar alarm, intercom system, or carry an audio signal.

After you have determined what services are being carried, try to draw a diagram similar to Figure B-1 that shows all existing telephone wiring in the house. Ideally, you need to know where all the jacks (or equivalent junction boxes if you still have some phones that don't have jacks) are and how the existing wiring gets to them. The actual path that it takes is not critical. The important issue is trying to determine whether or not there is anything unknown along the way. It is the hidden things that might cause trouble. An unused jack wired in the path probably won't cause any trouble unless someone inadvertently plugs a telephone or some other piece of equipment into it. Similarly, an unknown bridged tap may or may not be a problem. Knowing about those types of things ahead of time lets you deal with them, if necessary, instead of being surprised by them.

ISDN to a Single Target Room (Using Existing Wiring)

Use this section to help guide your ISDN wiring activities if:

- you already have at least one POTS lines and will be adding ISDN to an unused pair
- you have two or more POTS lines and you will be converting one of the existing POTS lines to ISDN service
- you want to have U-interface access to a service outlet

The goal is to use the existing customer premises wiring to serve the new ISDN line. In most cases, these procedures will provide satisfactory service because your premises is currently wired with quad or UTP wire. Due to the wide diversity of wire that is installed in North America today, the procedures described herein may not result in a successful installation. If it is determined the existing wiring cannot serve the new ISDN line, then you will be directed to our discussion on *ISDN to a Single Target Room*, which will explain how to run new wiring.

The section is written assuming the following:

- The building is a single-family home or small business (ground-level business).
- The existing service to the building is POTS.
- The ISDN line is only going to one room in the building.
- There is already a telephone outlet in the room where ISDN desired.
- The existing premises wiring is either quad cable or UTP.

The target configuration for this section would be that of residence or ground-level business that provides accommodations for office functionally. In addition to this home office, basic services provided by POTS are assumed to be spread well throughout the residence.

Installation procedures

This section will step you through the tasks involved in adding an ISDN line to your residence when you intend to retain your POTS service, and you have a spare pair in your wiring to support the new ISDN service. You may encounter slightly different situations than what the following tasks describe.

Task 1 - Locate the demarc device

To have the ISDN U-interface wired in a specific room using the existing premises wiring, it is necessary to survey the pairs of the customer wiring to find a spare pair of wires. First, you will need to identify an electrically isolated pair that has continuity from the demarc to the target room. Therefore, you will need to find the demarc. There should be one demarc for each line going into your building. It can be found either inside or just outside the walls of the building. Generally, there are two types of demarcs you might find. As stated earlier in this Appendix, only Telco personnel may make changes to this type of demarc directly.

It is important to understand that, in the case of an older type of demarc device, there is no legal point of order to modify the point of demarcation between the Telco network and the inside wiring. In order to modify the wiring that is connected to this type of demarc, a legal point of demarcation must be created. This can be done by using an additional service outlet or entrance bridge to connect inside wiring to. These configurations are depicted in Figure B-13 . The conversion process to add either a service outlet or entrance bridge to a legal demarc can be found at the start of this Appendix.

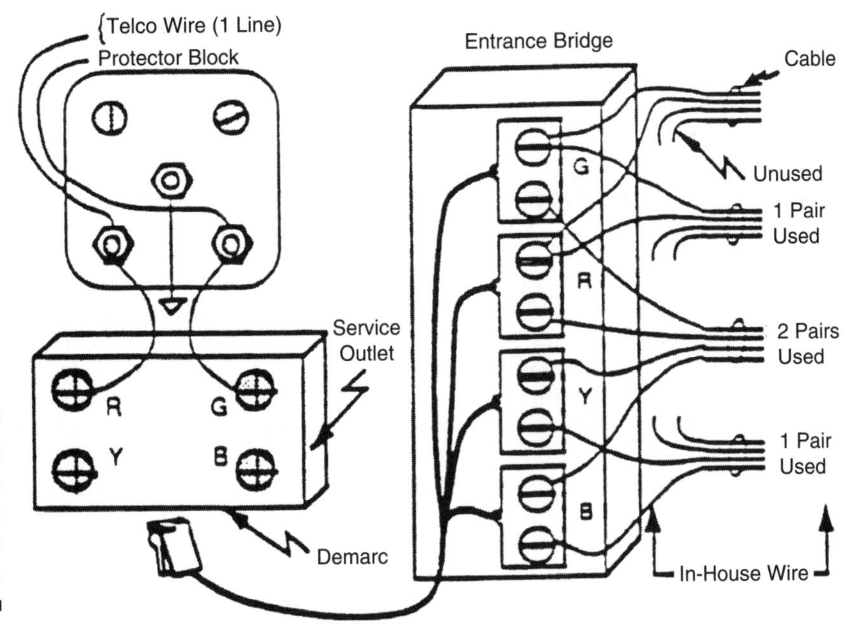

Figure B-13: Older-type demarc with entrance bridge.

Note: As you can see, there are screws and bolts on this demarc that can be adjusted. Only telephone company personnel may make any changes to the demarc device.

Otherwise, the demarc could resemble the one depicted in Figure B-14.

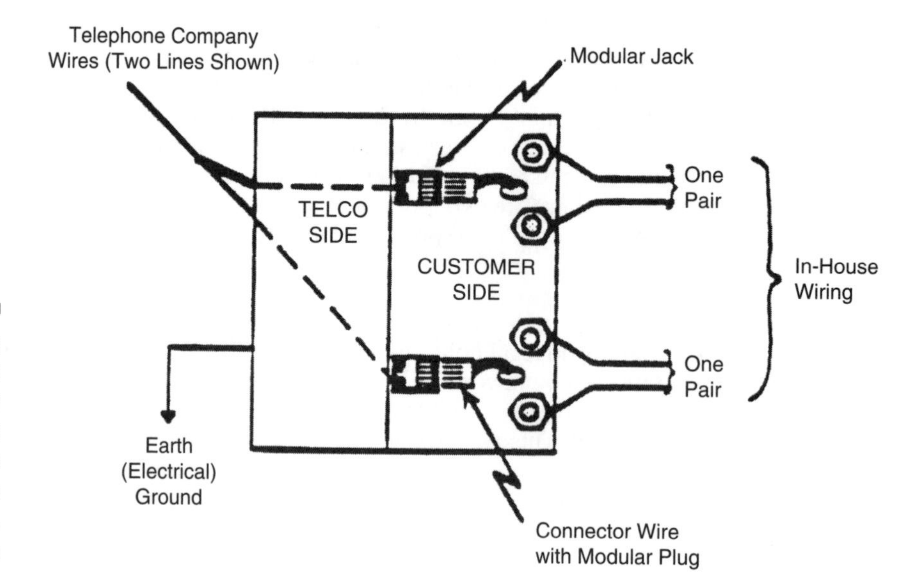

Figure B-14: Typical modular demarc device (new version).

This device is made with a short, modular pigtail so that the customer can plug in a telephone or terminal equipment to isolate any problems that may be found with the line.

Task 2 - Identify unused wire pairs at the demarc

Once the POTS demarc has been located, you should be able to find the customer premises wiring. Typically, the sheath of the customer premises wiring is gray, white, or beige and the sheath contains several colored conductors. Sketch and label the pairs that are connected to the demarc. For safety, disconnect the modular pigtail to separate the customer premises wiring from the network. For the older-type demarc, disconnect all of the wires connected to the customer premises wiring side of the demarc (see Figure B-13).

Most homes have customer wiring that is composed of quad cable that has two pairs within it. Newer installations could be using UTP cable that has many pairs within.

If only one pair is found in the customer premises wiring, then there are no spare pairs. In order to install the ISDN U-interface, proceed to the discussior of ISDN to the Single Target Room, which explains how to install new wiring for ISDN.

The conductors in telephone wiring are color coded. They may be a solid color or striped. Table B-2 depicts color combinations identifying cable pairs.

Table B-2: Color combinations identifying cable pairs

Pair Numbering	*quad cable Color Coding*	*T568A Color Coding*
Pair 1	Green and Red	White/Blue and Blue
Pair 2	Black and Yellow	White/Orange and Orange
Pair 3	White and Blue	White/Green and Green
Pair 4		White/Brown and Brown

Note: *For T568B color coding, Pairs 2 and 3 are swapped.*

Warning: As you connect ISDN or any telephone service, it is important to maintain the pairing of the wires. Do not cross-match wires from different color combinations. For example, green and yellow wires are never used together as a pair. ■

Identify which pairs of wires are in use at the demarc. If the demarc is modular, plug a POTS telephone set into each modular jack on the demarc. If you can hear dial tone when connected to a jack on the demarc, it indicates an active POTS line.

If the demarc is the older type and you have multiple demarcs (located within the same or different physical housings), then construct an adapter (see Figure B-15) using a surface-mount outlet, and verify live POTS service at all demarcs.

Any wire pairs connected to the live jack should be considered *inuse*. Any pairs that are not connected to the live jack should be considered potential spare pairs. Also, any pairs connected to a demarc jack that is not active (does not have dial tone) should be considered potential spare pairs. (Pairs that are connected to a demarc jack that has no dial tone should be disconnected.)

For the spare pairs identified at the demarc, make note of the colors of the conductors.

Warning: Some of the pairs that appear spare at the demarc may be in use for other purposes in the residence. You may need to check with the residents to learn if they have any knowledge of the customer premises wire being used for intercoms, stereo speakers, home alarms, local area network for computers, or similar applications within the residence. If so, they may be able to show you where it is used. Otherwise, you may choose to quickly survey the residence to determine if any other equipment is connected to the customer premises wire. If so, you should determine the colors of the pairs used, and eliminate these pairs from your potential spare pairs. ■

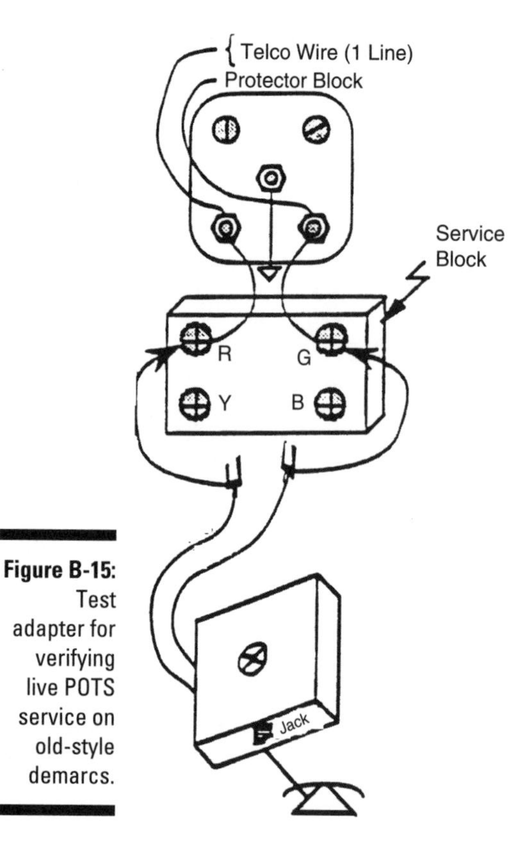

Figure B-15: Test adapter for verifying live POTS service on old-style demarcs.

Task 3 - Wiring the U-interface service outlet in the desired room

Locate the existing telephone-service outlet in the room where ISDN is desired. Open the service outlet, and determine if any of the pairs you have identified as spare appear to be present.

If the spare pair is not present at this service outlet, you will need to run new wiring to serve the ISDN service outlet. If there is no existing service outlet in the room, you will also need to run new wiring.

Once a pair of unused conductors is found at both the demarc and ISDN room, you may want to perform a continuity test on the spare pair that has been identified at the service outlet. This can be accomplished by twisting the conductors together at the demarc location and measuring the resistance of the pair at the service outlet with an ohm meter. If the continuity test fails, you may need to troubleshoot this problem further. For example, the premises may have been wired in a daisy chain configuration and the spare pair may not have been wired to each service outlet in the chain. The installer may want to review the section on New Wiring instructions before attempting this troubleshooting process. Installing new wire may be easier and less time consuming than attempting to correct a wiring deficiency.

If a continuity test is positive, a new service outlet for the ISDN access should be installed. Service outlets for POTS provide either four or six conductors, and actually use the center two pins for service. To wire the ISDN access, a six-position modular service outlet (sometimes referred to as an RJ-45 style service outlet) can be used. An eight-position modular service outlet is preferred. There are many types of service outlets and service outlet adapters that can be used to wire additional outlets instead of replacing or adding an additional service outlet. The following instructions will assume that the service outlet used for the ISDN access is placed directly next to the service outlet currently used for the first POTS line.

Mount the service outlet that will be used for ISDN access directly next to the existing service outlet. Be sure to make note of which service outlet is used for POTS and which service outlet will be used for ISDN. Remove the cover of both service outlets. Find the spare pair of conductors that you have decided to use for ISDN in the service outlet for POTS. Connect these two conductors to the two center pins of the new ISDN service outlet. When you make the connection of the spare pair to the ISDN service outlet, you can either, remove the spare pair from the POTS service outlet completely or wire between the two service outlets.

Note: The actual method of connecting the wires may be different for each service outlet, so be sure to read the manufacturer's instructions carefully. ■

The service outlet's terminals may be color coded as described in Table B-2.

After the ISDN service outlets have been wired, reattach the cover for both the POTS and the ISDN service outlets. Label the ISDN service outlet ISDN U with a marker. Also label the spare pair you have selected at the demarc.

Task 4 - Verify continuity at the ISDN service outlet

Before the ISDN service is connected, you may wish to verify the continuity between the spare pair you identified at the demarc and the pair you connected to the new ISDN service outlet. If your ISDN service is not working, you can use this procedure to troubleshoot the wiring between the service outlet and the demarc.

Note: This task requires you to disconnect any possible add several connections to your demarc(s). Before starting this process, you may wish to draw a sketch of the current wiring that is in place at the demarc(s). After performing Tasks 4 and 5, you will be instructed to reconnect your wiring in exactly the same way that you depicted in your sketch. ■

Disconnect any ISDN equipment from the ISDN service outlet. At the demarc, connect the pair your selected for ISDN to the pair currently used to run POTS to other service outlets in your premises. This can be done in a number of ways depending on the type of demarc you have.

If you followed the instructions in Task 2, you have already disconnected the wires attached to the customer premises side of the POTS demarc (if you have not performed this function, do it now). Using a pair of alligator clip leads, connect the customer side of the POTS demarc to the customer side of the ISDN demarc. (If you have a modular demarc that supports multiple pairs, the procedure should be as easy as unplugging each of the pigtails from the POTS and ISDN modular jacks. Leave the POTS pigtail disconnected and plug the ISDN pigtail into the POTS modular plug. Or, if the pigtails are not long enough, the alligator clip leads can be connected between the terminal lugs of the respective pairs.) Now, plug a POTS telephone into the ISDN service outlet. If you hear dial tone through the phone, you have verified the continuity of the wiring from the service outlet to the demarc. If you do not hear dial tone, there is a problem in the wiring, and you will need to select a different spare pair between the demarc and the ISDN service outlet. If no more spare pairs exist, you will need to run new wiring.

Note: Proceed with the following task before disconnecting your jumper wires. ■

Task 5 - Block other jacks using same conductor

You may wish to test the customer premises wiring to determine if any other jacks in the residence are connected to the ISDN pair, in addition to the ISDN jack you installed. If other jacks are connected to the ISDN pair, you may wish to plug those phone jacks with dummy plugs and label the jacks, to prevent equipment from being connected to those jacks. You should not disconnect any jacks you find connected to the ISDN pair. This could break the continuity to the ISDN jack, and your ISDN service would not work. With the jumpered connections left in place from the previous task, perform the following: Take a POTS telephone set to every other telephone jack in the residence, and plug it in and listen for dial tone. Any jack at which you can hear dial tone is also connected to the ISDN pair. Once you have tested all the jacks in the residence, make sure you have disconnected the POTS phone you were using for your testing. Return to the demarc, and reconnect the POTS and ISDN demarcs as they are depicted in the sketch you made in Task 4.

Task 6 - Verify that POTS is OK

You now need to check to see that the POTS access was not indirectly affected. Reconnect the pigtail you identified for POTS at the demarc that you disconnected as part of Task 2 (or reconnect the wires to the bolt down terminals if you have an old-type demarc). Return to the room where the new line has been installed. Connect a standard telephone to the POTS jack and pick up the handset. A dial tone should be heard, possibly after a few minutes delay. If it is not heard, recheck the connections in the POTS jack and at the demarc. Now that the POTS line has been reconnected at the demarc, all other telephone equipment connected to the original POTS line should also be working again. If it is not, you will need to troubleshoot the problem.

Task 7 - ISDN service connection

Verify that the service provider has connected and activated your ISDN service. You may now connect the pair of wires you selected for ISDN service in the customer premises wiring to the appropriate ISDN access at the demarc. The service provider may have identified the connection points for ISDN on the demarc device. The identification will often be a notation of the phone number assigned to your ISDN line next to one of the demarc connecting points. Connect the ISDN pair you selected and tested to the terminals of the demarc identified for your ISDN service. Be sure the pigtail (if one exists) for the ISDN demarc is plugged in.

Task 8 - Connecting your ISDN equipment

At the ISDN jack you installed you may now connect your ISDN equipment. A typical arrangement includes a power supply, and NT1 device, and an ISDN telephone. Another configuration includes the use of multiple pieces of terminal. The wiring topology that should be utilized in wiring multiple terminals on the S/T bus is the star wiring configuration.

Task 9 - Connecting your ISDN NT1 equipment

Proper selection of NT1 timing and terminating resistors is very important for satisfactory ISDN service. In most cases, where all of the ISDN terminal equipments are located within the same room (less than 250 feet per leg of the star), the NT1 should be configured so that it provides a 50-ohm resistance termination. This 50-ohm termination can be provided in many ways depending upon the design of the NT1. Some vendors provide a 50-ohm terminating resistor, while others provide a combination of terminators that achieves a 50-ohm termination. See the manufacturer's instructions for proper optioning of your specific piece of equipment.

If your NT1 has an option for fixed or adaptive timing, choose the default timing option equipped with your NT1. See the manufacturer's instructions for additional optioning information of your specific piece of equipment.

Plug the NT1 power supply into a wall outlet and follow the instructions provided to you from the manufacturer. You may need another cord with modular connectors to connect the power supply to the NT1 device. Again, consult the power supply and NT1 device instructions on the type of cable and where it should be connected on each device. Once the NT1 is connected to the power supply, you may notice indicators lighting to show the NT1 and ISDN line status.

Task 10 - Connecting your ISDN Terminal Equipment

The user should connect the ISDN terminal equipment directly to the NT1 that supports the star wiring configuration. Adapters may be required for connection of multiple terminals in a star configuration. The instructions for your ISDN terminal equipment will provide you with information regarding cables, power supplies, and (possible) terminating resistor options. In the wiring configuration discussed here, the proper option for termination resistors in the ISDN terminal equipment is *off or none.* Your terminal equipment may provide you with a straight through eight-conductor cord suitable for ISDN with eight-pin modular plugs at both ends to connect the NT1 to the ISDN terminal equipment. Plug the cable into the NT1 and ISDN terminal equipment as directed in the instructions that came with the NT1 and the terminal equipment. Program the required operating parameters in the ISDN set (such as SPID, Directory Number, etc.), following the manufacturer's instructions. Certain parameters (SPID, DNs) are provided by your service provider. If your ISDN service includes voice capabilities, you should now be able to make a voice call from you ISDN terminal equipment. If everything is operational, the installation is complete.

Task 11 - Troubleshooting

If your ISDN service is not working as you expect, you may need to troubleshoot your configuration:

- Is everything plugged in as it is supposed to be?
- Do status indicators on the NT1 and/or ISDN telephone appear OK, based on instructions that came with the equipment?
- Is your ISDN telephone programmed with the correct parameters?
- Are the terminating resistors optioned correctly in the NT1 (50-ohms) and the ISDN terminal equipment (off or none)?

You may use the procedure in the section titled *Task-Verify continuity for the New ISDN Jack* to determine if there is a wiring problem to the ISDN jack. To verify that your ISDN service from your service provider is working correctly, you may connect the ISDN equipment directly to the test jack at the demarc by disconnecting the pigtail and wiring the NT1, ISDN, and necessary power supplies to the demarc. Then attempt to access ISDN service.

Presence of a distribution device

Some residences include a distribution device. This is a connecting block that can be used to fan out the customer premises wiring from the demarc to the rest of the residence.

The distribution device, if used, will often be located inside the residence, near the area where the wiring from the demarc (on the outside of the residence) enters the structure. It will have terminals for connecting the wires leading to the demarc, and for connecting the customer premises wiring leading to the rest of the residence. If a distribution device is present, the task for locating a spare pair between the demarc and ISDN jack will need to be performed in two steps. You will need to locate a spare pair between the demarc and the distribution device, and then you will need to locate another spare pair between the distribution device and the ISDN jack.

Distribution devices can be used in other locations within a residence. For example, if there are multiple telephone jacks in a room, there may be a distribution device within the room to fan out the wiring to multiple jacks.

Multiple POTS Line Session

The tasks for adding ISDN to a residence with multiple POTS lines, when one spare pair already exists or one POTS line will be changed to ISDN service, are similar to those described in Installation Procedures. As was previously mentioned, the goal is to incorporate (when possible) the use of existing customer premises wiring to serve the new ISDN line. In scenarios where the premises has multiple POTS lines in use, a few possibilities exist. For example, if there is a spare pair contained within the premises wiring, the installer should follow the tasks provided in the Installation Procedures.

If there is not a spare pair in the wiring, then the installer has a choice. Either the installer can add customer premises wiring to the premises by following the instructions in *ISDN to a Single Target Room* section, or the installer could use a pair of wires that a current POTS phone is currently using for ISDN service. If this POTS phone is an extension line to another phone in the premises, then the installer would have to trace the pair of wires used to provide the extension service and disconnect them appropriately. Then, the Installation Procedure tasks can be followed. If this POTS phone is the only phone tied to a specific POTS demarc, then the installer would need to call the service provider, cancel POTS service to that demarc, and specify ISDN service to that demarc. The installer could then follow the instructions cited in Task 7.

ISDN to a Single Target Room, New Wiring

This section explains how to install ISDN in the residence or small business when the existing wiring does not have a spare pair or will not support ISDN for other reasons. You might want new wiring to prepare for future services or because the existing wiring quality is suspect. This section will help you understand the relevant standards and products you will install. If your house already meets EIA/TIA 570 standards, then read this section to understand how it supports ISDN in a single target room.

Wiring standards

The reason for discussing standards is to increase awareness and promote adherence in practice. This section contains a summary of the wiring standard applicable to ISDN. Refer to EIA/TIA 570 Residential and Light Commercial

Telecommunication Wiring Standard for more details. This standard specifies the number of conductors and cable quality. Cables for new installations should contain four twisted-pairs of solid conductors. The additional pairs will support new functions, such as DC power distribution to terminal equipments. The standard recommends eight-position modular jacks with eight conductors. The pinpoint for these jacks (called T568tA or T567B) can be found in Table B-1. Choose either T568A or T568B jacks and use throughout the building. The TSB-40 standard defines five levels of categories of cable characteristics. Category 3 cable meets the requirements for new installation of telephone and ISDN services. Installing Category 4 or Category 5 cable may save money in the long run if future higher-rate services are expected. The standards specify star wiring to each service outlet from the distribution device. Although daisy chaining outlets in the same room is allowed by the standard, it is not recommended. The length of cable from the distribution device to each service outlet shall not exceed 90 meters (295 feet).

Warning: We do not recommend that wiring leave the building. If it does, consult your local codes for additional protection requirements. Cable exposed to the elements may degrade. ■

Distribution device

The distribution device required by EIA/TIA 570 allows multiple outputs for a single input. Wires placed over insulation displacement connectors in multiples of four make connections by snapping a cap onto the connector. No special tools are needed. This device can support star wiring (multiple outputs for a single input) to multiple target rooms by connecting the ISDN signal to several input rows. For ISDN in one target room, only one output is needed.

Modular jacks

When installing four-pair cables, the modular jacks must terminate all pairs. The acceptable modular jacks conform to T568A or T568B wiring. These eight-position modular jacks contain eight conductors. EIA/TIA categories also apply to modular jacks. The minimum requirement is Category 3 jacks when rewiring or in new installations. The standard recommends matching the cable and jack categories, a Category 3 jack used with a higher-Category cable provides improved performance over a Category 3 cable for ISDN and allows for easier system upgrade.

Modular jacks for ISDN differ from the typical POTS jack. Color code or label the jack with ISDN U to eliminate confusion and prevent connecting incompatible equipment to the jack. The Category 3 connecting cords used are also different from ordinary phone cord.

NT1 in the target room

The NT1 and a power source will also reside in the target room. Some supplies can power the ISDN terminal equipment as well. Only one pair is used to carry the ISDN U signal to the target room. However, in the future you may want to expand ISDN to more target rooms. Then, the four-pair cable and distribution device are essential for simple and inexpensive rearrangement. Centrally locate the distribution device for the possible future conversion to multiple target rooms. Note that some terminal equipment may have an integrated NT1. This type of equipment may preclude growth to multiple ISDN terminals.

Wiring summary

The following is a summary of minimum equipment requirements for rewiring according to the standards.

- Category 3 or higher cables and suitable connecting cords
- Connecting cords should not exceed 10 feet
- Category 3 or higher modular jack intended for T568A or T568B wiring
- A distribution device (not essential for this application, but required for full standards compliance and recommended for flexibility)

The following tasks are necessary to install/verify the ISDN service from the demarc device to the target room:

Task 12 - Wiring from the demarc to the target room

Plan the route from the demarc to the target room to determine the length of the cable needed. To avoid shock, do not connect the cable to the legal demarc yet. If you are replacing an existing service, you may not have a legal demarc point. In this case, refer to POTS Components section to create a legal demarc. When fastening cable to the building take care not to crush or puncture the cable.

Task 13 - Using a distribution device

Mount a distribution device in a suitable location, preferably centralized to minimize the wire runs for future wiring. Even though the ISDN service uses one pair, connect all four pairs to the distribution device if it accepts four pairs, according to the color code in Table B-1. Run the cable to the target room, leaving some slack at the distribution device.

Task 14 - No distribution device (not recommended)

Since you have a single target room, a distribution device is not really needed. However, if you want to move the service to another room or support the future conversion to multiple rooms, the distribution device allows the flexibility for simple changes. If you choose this method, start at the legal demarc, leave some slack, and run the cable to the target room.

Task 15 - Installing the 8 pin modular jack

Once the wiring reaches the desired location, strip back the outer jacket and connect all four pairs to the eight-pin modular jack according to the color code in Table B-1. Mount the jack to the wall or baseboard and place the cover plate on the hack. You should now have a continuous pair from the demarc to the target outlet. Label this outlet ISDN U to avoid connecting the wrong equipment. Now connect pair one of the cable coming from the distribution device (if present) to the ISDN service at the legal demarc.

Task 16 - Connecting equipment in the target room

Now connect your equipment to the ISDN jack. (This is only one way to connect equipment; other power and NT1 arrangements may exist. Consult the guidelines from your equipment manufacturer.)

The NT1 should provide a 50-ohm resistance termination. This 50-ohm termination can be provided in many ways depending on the NT1 design. Some vendors provide a 50-ohm terminating resistor, others may provide a combination of terminators that achieves a 50-ohm termination. See the manufacturer's instructions for proper NT1 optioning.

If your NT1 has a timing option, choose the default.

Connect the power supply to the ISDN modular jack and the NT1 to the power supply. Then connect the ISDN terminal to the NT1. Use Category 3 or better cords. Plug the power supply into an AC outlet. Program your ISDN terminal equipment with the appropriate parameters (DNs, SPID, etc.) following the manufacturer's instructions. Certain parameters (SPID, DNs) are provided by your service provider. If your service includes voice, your should now be able to make a call from your ISDN terminal equipment. If you ordered data service, then the terminal instructions should guide you through options needed to establish data calls. Once the power supply is plugged into AC, you may notice indicators lighting on the NT1 that show line and equipment status.

Troubleshooting

If your ISDN service does not work, you may need to troubleshoot your installation.

- Is everything plugged in correctly?
- Do status indicators appear OK based on instructions that came with the equipment?
- Are the terminating resistors correct in the NT1 (50 ohms) and the terminal (off or none)?
- Does the ISDN equipment pass its own self-test?
- Is the problem eliminated by connecting your equipment (power, NT1 and phone) at the legal demarc. If the service works there, then recheck your wiring.

Common wiring problems include transposed wires or pairs, open or continuity failure, and shortened conductors. When checking for continuity, begin at the legal demarc and test through the ISDN jack. Many wiring errors occur where cable connections are made, at the distribution device, the wall jack, or the legal demarc.

Glossary

analog communications The method of voice transmission used in today's telephone system. This method converts voice to electrical signals and amplifies them so the voice can be sent over long distances. Using analog for data transmission is at the bottom of the telecommunications bandwidth food chain.

ANSI American National Standards Institute. The primary standards organization for the U.S., ANSI plays a significant role in defining ISDN standards.

asynchronous communications The form of data communications that transmits data one character at a time with start and stop bits. The method used for data communications over POTS using modems.

ATM Asynchronous Transfer Mode. A form of fast packet switching that allows for data transmission via Broadband ISDN, a faster form of digital communications than ISDN.

AT&T 5ESS The leading telephone switch platform, made by AT&T. These switches use Custom (proprietary) or NI-1 software.

B channel Bearer channel. A 64 Kbps bearer channel used for delivering data or voice communications over ISDN. The standard BRI connection includes two B channels, for a total uncompressed capacity of 128 Kbps.

bandwidth The amount of data that can flow through a channel. The greater the bandwidth, the more data that can travel at one time.

bearer services A communication connection's capability to carry voice, circuit, or packet data. The two B channels in a BRI connection are bearer channels.

Bellcore Bell Communications Research. The research arm of the RBOCs. Bellcore was part of Bell Laboratories before the breakup of AT&T. Bellcore plays a leading role in developing ISDN standards and other ISDN activities among its member telephone companies.

BONDING Bandwidth on Demand Interoperability Group. The concatenating of two or more B channels to form a single channel with a bandwidth greater than 64 Kbps. For applications such as desktop video conferencing, bonding combines the two B channels for a total of 128 Kbps to transmit video and audio.

bps Bits per second. The unit of measurement for data transmission speed over a data communications line.

BRI Basic Rate Interface. A defined interface to ISDN that includes two B (bearer) channels and one D (data) channel. Commonly referred to as 2B+D. Other configuration options are available within the BRI interface, depending on your telephone company. For example, you can get only one B channel, or two B channels and no D channel.

bridge A device that connects together two networks of the same type.

BISDN Broadband ISDN. A type of ISDN service that uses fiber-optic lines and ATM to deliver bearer services with data transmission rates of more than 150 Mbps. This is the next generation of ISDN service that will replace ISDN delivered via the copper wiring used today.

CACH EKTS Call Appearance Call Handling Electronic Key Telephone Service. Supplements EKTS to allow more than one directory number and multiple call appearances on each directory number.

call appearances A supplementary ISDN service that allows multiple incoming calls. Each directory number can have multiple call appearances, depending on the switch type.

Caller ID A telephone company service that delivers the calling party's telephone number to the called party, which can appear on an ISDN telephone, an LCD screen, a computer screen, or on another device.

CCITT Comité Consultatif International de Télégraphie, or International Telephone and Telegraph Consultative Committee. This organization is now called ITU, which stands for International Telecommunications Union. A United Nations organization that produces recommendations for standards for international ISDN.

Centrex A service offering by LECs that provides local switching features similar to those provided by an on-site PBX.

circuit-switched data Data sent over a circuit-switching network, of which ISDN is one.

circuit switching A form of communication in which an information transmission path between two devices is routed through one or more switches. The path is assigned for the duration of a call.

client/server computing The foundation for networking, in which one computer acts as the host or server, and the other computer acts as a client. In the case of remote access, your PC acts as the client computer that connects to a server.

cloud A commonly-used term that defines any large network, such as ISDN.

CO Central office. The site where the local telephone switches reside for all the telephone system's call routing and other functions. This is the telephone company side of the local loop.

CODEC coder/decoder. Transforms analog data into a digital data form and converts digital data back to analog form.

common carrier Telephone companies that provide long-distance telecommunication services, such as AT&T and MCI.

compression A process for reducing the number of bits required to transmit information. For ISDN, the result of compression is data transmission speeds up to four times faster than without compression, or about 512 Kbps.

Corporation for Open Systems COS, a member-based organization that promotes open systems and connectivity. COS is instrumental in developing ISDN Ordering Codes for streamlining the acquisition of ISDN service from the telephone companies. COS is also instrumental in getting industry support for ISDN-1 standards.

CPE Consumer-provided equipment. The equipment after the point at which the telephone company terminates the line to the premises. In the U.S., CPE includes the NT1 device. End users must purchase or lease, install, and maintain their own CPE. In many other parts of the world, end-user equipment is part of the service provided by the telephone company.

CSD Circuit-Switched Data. An ISDN circuit switched call for data in which a transmission path between two users is assigned for the duration of a call at a constant, fixed rate.

CSV Circuit-Switched Voice. An ISDN circuit-switched call for voice in which the transmission path between two users is assigned for the duration of a call at a constant, fixed rate.

CSV/CSD Alternate Circuit-Switched Voice/Circuit-Switched Data. A B channel configuration that allows either circuit switch voice or data communication.

D channel Data channel. The separate channel for out-of-band signaling between the user and the ISDN network. The D channel can also be used to deliver X.25 data packets at up to 16 Kbps.

demarcation point The point at the customer premises where the line from the telephone company meets the premises wiring. From the demarcation point, the end user is responsible for the wiring. The physical device that provides the means to connect the telephone company's wire to the premises wiring is called a network interface box.

desktop video conferencing A PC-based video conferencing system that allows people to conduct video conferencing in real time from their desks. The basic desktop video conferencing system includes a video camera, a video card, and an ISDN adapter card.

DN Directory number. Each BRI connection can have up to two directory numbers, one for each B channel. Directory numbers are telephone numbers for ISDN.

DSS1 Digital Subscriber Signaling System No. 1. The network access signaling protocol for users connecting to ISDN. It includes the CCITT Q.931 and Q.932 standards.

DTE Data Terminal Equipment. Any device that converts information into digital signals for transmission or reconverts digital information into another form.

EKTS Electronic Key Telephone Service. The National ISDN-1 standard for working with supplementary services on an ISDN telephone or analog telephone connected to an NT1 Plus device.

Ethernet The local area network protocol used in most PC networks. Typically, most Ethernet networks support data transmission speeds up to 10 Mbps.

exchange area A geographical area in which a single, uniform set of tariffs for telephone service is in place. A call between any two points in an exchange area is considered a local call.

FCC Federal Communications Commission. The U.S. governmental agency responsible for regulating the telephone industry.

fiber optics A new generation of telecommunication wiring that uses light beams sent through thin strands of glass or other transparent materials. Fiber optics can transmit large amounts of data and form the physical transmission foundation for Broadband ISDN.

full duplex The bidirectional communication capability in which transmissions travel in both directions simultaneously.

functional devices A classification of ISDN operational functions used to describe what tasks different components of an ISDN configuration perform. For example, the Network Termination 1 function defines the NT1 device that presents your premises as a node on the ISDN network. Another functional device is the terminal adapter, which defines the role of an adapter to convert some other form of communication to ISDN. For example, a TA allows an analog telephone to communicate over an ISDN device.

Group 3 Fax Currently, the most widely-used facsimile protocol, which operates over analog telephone lines or with a terminal adapter over ISDN.

Group 4 Fax A facsimile protocol that allows high-speed, digital fax machines to operate over ISDN.

half duplex Data transmission that takes place in only one direction at a time.

IEC InterExchange Carrier. The telephone company that provides telephone service outside the local telephone companies. For example, AT&T and MCI are InterExchange Carriers. InterExchange Carriers are also referred to as common carriers.

IEEE 803.2 The protocol that defines an Ethernet network at the physical layer of network signaling and cabling.

in-band signaling Network signaling that is carried in the same channel as the bearer traffic. In analog telephone communication, the same circuits used to carry voice are used to transmit the signal for the telephone network. Touch Tone signals are an example of in-band signaling.

interface A specification that defines the protocols used at a particular reference point in a network. The Basic Rate Interface (BRI) refers to an access interface to ISDN.

Internet A huge global network of networks based on the TCP/IP suite of protocols. The Internet currently encompasses more than 15,000 networks connecting more than 30 million users and growing rapidly.

internetworking Data communication across different network operating systems.

interoperable Two pieces of equipment are interoperable when they work together. Standards make devices from different vendors work with each other. For example, the H.320 standard for video conferencing allows you to use an Intel ProShare system to connect to a Vivo320 system.

IP Internet Protocol. The internetworking protocol that forms the basis of the Internet.

IPX Internet Packet Exchange. Novell's NetWare internetworking protocols.

ISDN Integrated Services Digital Network. The next generation of telecommunication services offered by telephone companies. ISDN delivers digital communication via standard POTS lines at a speed of 128 Kbps without compression.

ISDN address The address of a specific ISDN device. It comprises an ISDN number plus additional digits that identify a specific terminal at a user's interface. An ISDN number is the network address associated with a user's ISDN connection.

ISDN telephone A telephone designed for ISDN service. It typically includes programmable buttons for managing call features and an LCD display for viewing caller information.

ITC Independent Telephone Company. In the U.S., a telephone company that was not owned by AT&T prior to divestiture.

ITU International Telecommunications Union. An organization under the United Nations that prepares telecommunication recommendations or standards, including many related to ISDN. The ITU was formally the CCITT.

Kbps Kilobits per second. The unit of measurement in thousands of bits per second for data transmission. ISDN has a data transmission capacity of 128 Kbps for the two B channels, or 64 Kbps for each B channel.

key systems Telephone equipment with extra buttons that provide users with more functionality than regular telephones. ISDN phones and NT1 Plus devices that support analog telephones include key systems. A key system is a protocol invoked when you press a sequence of keys on the analog or ISDN telephone's dialing pad.

LAN Local area network. A group of computer and other devices linked via a network operating system. LANs vary in size but are restricted to a single location because of cabling limitations. The leading protocol for LANs is Ethernet. Leading PC LANs include Novell's NetWare, Windows for Workgroups, and Windows NT.

LATA Local Access Transport Area. Local exchange carriers (RBOCs) provide service within a LATA. Typically, a LATA comprises multiple area codes. In most cases RBOCs are prohibited from offering telecommunication services between LATAs.

LEC Local exchange carrier. The local telephone company. An LEC is also called an RBOC (Regional Bell Operating Company).

local loop The pair of copper wires that connects the end user to the telephone company's central office, which is the gateway to the global telephone network. These wires, originally installed for analog communication, are the same wires used for ISDN service but require new equipment at the end user's premises and at the telephone company.

logical channels The three channels of a BRI connection, which are defined not as three physically separate wires but as three separate ISDN system channels.

LT Line termination. Defines the local loop at the telephone company side of an ISDN connection to match the NT1 function at the customer end of the local loop.

Mbps Million bits per second. A measurement for high-speed data transmission used in LANs and other digital communication links.

modem Modulator/demodulator. A device used to send data over analog telephone lines. It converts digital signals to analog signals at the sending end and converts analog signals to digital signals at the receiving end.

NANP North American Numbering Plan. The familiar ten-digit numbering system used today in the U.S., Canada, and Mexico, which includes the three-digit area code followed by the seven-digit local telephone number.

National ISDN Defined by Bellcore, National ISDN 1 (NI-1) is an agreement among telephone companies and CPE vendors to jointly provide the first phase of standards-based ISDN. NI-1 is a collection of standards to allow CPE to work across different telephone company switches using the Basic Rate Interface.

NDIS Network Driver Interface Specification. Developed by Microsoft, NDIS provides a common set of rules for network adapter manufacturers and network operating system (NOS) developers to use for communication between the network adapter and the NOS. Most network adapters now ship with an NDIS driver. If the NOS you use supports NDIS, which most do, you can use any network adapter that has an NDIS driver.

NetBEUI NetBIOS Extended User Interface. Microsoft's implementation of NetBIOS used in Windows for Workgroups.

NetBIOS Developed by IBM and used in DOS-based networks.

network interface box The point where the lines from the telephone company meet the wiring for your premises.

NIUF North American ISDN User's Forum. The National Institute for Standards and Technology formed the NIUF in conjunction with ISDN industry players to identify ISDN applications and to encourage ISDN equipment vendors to develop CPE to meet end-user needs.

NT DMS-100 A telephone switch platform made by Northern Telecom. These switches use proprietary or NI-1 software.

NT1 Network Termination 1. Located at the end-user side of the ISDN connection, this functional device represents the termination of the ISDN system at the end user's location. The NT1 function is embedded in NT1 and NT1 Plus devices.

NT1 Plus device A device that includes a built-in NT1 as well as ports to connect other devices (analog, ISDN, or X.25) to an ISDN line.

NT2 Network Termination 2. A device that handles network termination and switching functions, typically embodied in PBXs (Private Branch Exchanges). An NT2 device performs intelligent operations such as switching and concentrating traffic across multiple B channels in a PRI line.

ODI Open Datalink Interface. The specification developed by Novell for supporting different adapters and network operating systems (NOS).

out-of-band signaling Allows telephone network management signaling functions and other services to be sent over a separate channel rather than the bearer channel. ISDN uses out-of-band signaling via the D channel. Out-of-band signaling used in ISDN consists of messages rather than audio signals, as is the case with the Touch Tone analog telephone system.

packet switching A data transmission method in which data are transferred via packets. A packet is a block of data. Packets are sent using a store-and-forward method across nodes in a network.

passband The frequency spectrum that determines the amount of data that can be transmitted through a channel. The passband is what determines the bandwidth of a channel.

passive bus Refers to the ability to connect multiple devices to a single BRI connection without repeaters to boost the signal. The configuration of the passive bus combines the terminating resistance for all the devices connected to your ISDN line to add up to 100 ohms.

PBX Private Branch Exchange. A telecommunication switch at a customer's premises that handles call management. The PBX connects to the telephone company via a dedicated, high-speed communications link to transport a large volume of traffic. Typically, PBX systems handle the internal telecommunication needs of large organizations.

PCM pulse code modulation. The method used to convert analog audio to digitized audio for ISDN.

phantom power The ability of the NT1 to provide power to the Terminal Equipment 1 or terminal adapters via two wires in an eight-wire cable.

point-to-multipoint configuration A physical connection in which a single network termination supports multiple terminal equipment devices. This configuration is supported by the S/T-interface.

point-to-multipoint connection A connection established between one device on one end and more than one device on the other end.

point-to-point connection A connection established between two devices through ISDN.

point-to-point configuration A physical connection in which a single NT1 functional device supports only one device.

POS Point of sale. Any device used for handling transactions, such as card readers for credit card or debit card transactions.

POTS Plain Old Telephone Service. A term used for standard analog telecommunication.

powering The powering of the ISDN line and CPE equipment. The ISDN line and any CPE connected to it must be powered locally. Usually, these powering capabilities are built into the NT1 or NT1 Plus device.

PPP Point-to-Point Protocol. A communication protocol that allows a computer using TCP/IP to connect directly to the Internet. The new PPP/MP protocol is an improved version of this protocol for ISDN connection to the Internet.

PPP/MP Point-to-Point Protocol/Multilink Protocol. The new Point-to-Point Protocol for ISDN connection that allows use of both B channels for remote access to the Internet. PPP/MP also allows different remote access devices to communicate with each other.

PRI Primary Rate Interface. An ISDN interface designed for high-volume data communication. PRI consists of 23 B channels at 64 Kbps each and one 64 Kbps D channel.

protector block The point where the lines from the telephone company meet the lines from premises wiring before the network interface box.

protocol A set of rules that define how different computer systems and other devices interoperate with each other.

PS/2 Powering for any ISDN devices connected to an NT1 or NT1 Plus device that don't have a local power source.

PSC Public service commission. See PUC.

PSTN Public Switched Telephone Networks. A POTS-based system that uses analog signals between the branch exchange and each end device, such as a telephone or a modem.

PUC Public utilities commission. Also called public services commission (PSC). A governmental agency, usually at the state level, that regulates telephone companies and other utilities. PUCs define how you're charged for telephone services, which are called tariffs. The FCC deals with some similar functions at the federal level.

R-interface See R reference point.

R reference point The ISDN reference point that sits between the non-ISDN device and the terminal adapter (TA) functional device.

rate adaptation A system that allows two pieces of data equipment operating at different data transmission rates to interoperate.

RBOC Regional Bell Operating Company. One of the local exchange carriers that were created during the break up of AT&T. RBOCs provide telephone service in a region of the U.S. They currently can't offer long-distance telephone service between LATAs or manufacture equipment.

reference point A specific point in the model of how ISDN works. Each component of this model is identified using a reference point. For example, the U reference point defines the local loop of an ISDN connection. These reference points are also called interfaces, such as the U-interface or S/T-interface.

RJ-11 connector A modular connector used for four- or six-wire analog devices.

RJ-45 connector A modular jack that can hold up to four pairs of wires. It looks similar to an RJ-11 but is larger. ISDN connections use RJ-45 jacks at the S/T-interface.

RS-232 An industry standard for serial communication connections. The current version of this standard is RS-232C. Most PCs include one or more RS-232 ports for connecting devices, such as a modem and a mouse.

router A hardware device that acts as a gateway between different types of networks.

serial communication The transmission of data one bit at a time over a single line. Serial communication can be synchronous or asynchronous.

S reference point The ISDN reference point that represents where a CPE connects to a customer switching device, such as a PBX system. This type of device is called an NT2 functional device.

S/T-interface See S/T reference point.

S/T reference point Combines the ISDN reference points where a device connects to either an NT1 or NT2 functional device.

SS#7 Signaling System Number 7. A common channel signaling system that performs network signaling functions. Used to establish ISDN call functions.

SPID Service Profile Identification. An alphanumeric string that uniquely identifies the service capabilities of an ISDN terminal. This is an identifier that points to a particular location in the telephone company's central office switch memory where relevant details about the device are stored.

Standard A set of technical specifications used to establish uniformity in hardware and in software.

subscriber loop The pair of copper wires that connect the end user to the telephone network. These same wires are used to provide ISDN service but need the addition of the NT1 at the end-user location and the line termination at the central office.

Supplementary services The collection of voice communication services available via ISDN. These services include call management features such as call appearances, conference calling, and call forwarding.

Switch The equipment that connects users of the telecommunication network. Each subscriber has a dedicated loop to the nearest telephone switch. All of these switches have access to trunk lines for making calls beyond the local exchange area. A call from one user to another consists of a loop at each end of the connection, with switches and trunk lines used to route the connection between them.

synchronous communications A data transmission method in which data are transmitted in blocks separated by equal time intervals. This is a faster method of data communication than asynchronous, but both are serial communications.

T-interface See T reference point.

T reference point The ISDN reference point that represents where an ISDN device connects to an NT1 functional device.

TA Terminal adapter. The ISDN functional device that allows non-ISDN devices to work with ISDN.

tariff A rate and availability schedule for telecommunication services that is filed with and approved by a regulatory body to become effective. Tariffs also include general terms and conditions of service.

TCP/IP Transmission Control Protocol and Internet Protocol. The suite of networking protocols that let disparate types of computers communicate over the Internet.

TCP/IP stack The software that allows a computer to communicate via TCP/IP.

TE Terminal equipment. A term for any device connected to an ISDN line.

TE1 Terminal Equipment 1. Any ISDN-ready device that connects directly to ISDN. An ISDN telephone is an example of a TE1 device.

TE2 Terminal Equipment 2. Any non-ISDN device that must be used with a terminal adapter to work with ISDN. An analog telephone is an example of a TE2 device. The combination of a TE2 and a TA has the same functionally as a TE1.

telephony The marriage of computer and telecommunication.

TA Terminal adapter. Any device that adapts a non-ISDN terminal for an ISDN interface. A TA gives a TE2 device the functionality of a TE1 device.

U-interface See U reference point.

U reference point The reference point of an ISDN connection that includes the local loop wiring up to the NT1 functional device.

usage sensitive The cost of a service, such as ISDN or analog telephone service, that is based on the time you actually use the service.

UPS Uninterruptible power supply. A device that ensures a back-up power supply for electrical devices in the event of a power outage. For ISDN, NT1 Plus devices can include UPS for maintaining power for analog voice communication during a power outage.

whiteboard Collaboration software typically bundled with desktop video conferencing systems. It allows two users to share a computer screen in a similar manner to using a whiteboard in a meeting room.

WAN Wide area network. A communication network that connects geographically dispersed sites.

WinSock A program that conforms to a set of standards called the Windows Socket API (Application Programming Interface). A WinSock program controls the link between Microsoft Windows software and a TCP/IP program.

World Wide Web A hypertext multimedia-based system for accessing Internet resources. Commonly referred to as the Web or WWW, it lets users download files, listens to audio, and view images and videos. Users can jump around the Web using hyperlinks embedded in documents. The leading Web browsers are Netscape and Mosaic.

X.25 The protocol for packet-mode services as defined by CCITT. A CCITT interface standard that lets computing devices communicate via wide area packet-switched data networks.

Index

• Symbols •

• A •

• D •

• E •

• S •

• W •

Buy 1 "Vivo TeleWork-5" System and Get Your Second Kit For 1/2 Price

Take Advantage of this SPECIAL OFFER
Made Available Exclusively to Readers of
ISDN For Dummies™.

Buy Your First "Vivo TeleWork-5" System --
Get Your Second Kit for 1/2 Price
And Soon You Will Be:

1. Accessing the Internet at High-Speeds
2. Working off a Remote LAN Quickly and Easily
3. Videoconferencing with Colleagues All Around the World
4. Faxing Information Instantaneously
5. Modeming Documents in Seconds

USE YOUR ISDN LINE 5 DIFFERENT WAYS BY CALLING VIVO TODAY at 1-800-Vivo-411!!

Vivo Software, Inc. 411 Waverley Oaks Rd., Waltham, MA, 02154 Tel: 617-899-8900, Fax: 617-899-1400, E-mail: info@vivo.com

Internet address: http:\\world.std.com\~vivo. Vivo TeleWork-5 is a trademark of Vivo Software, Inc. (Pricing and Terms may change without notice.)

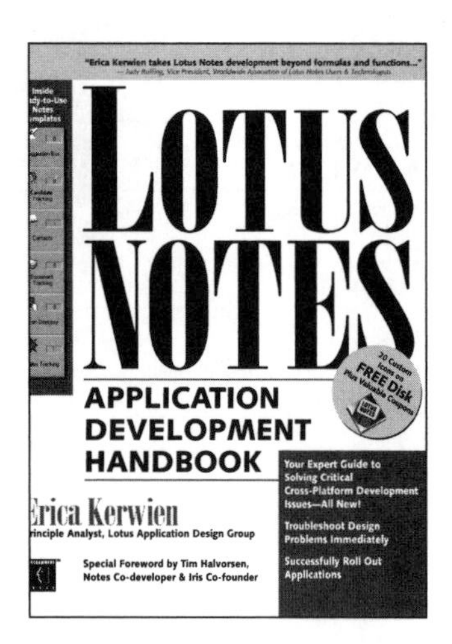

Lotus Notes Application Development Handbook
by Erica Kerwien

ISBN: 1-56884-308-9
$39.99 USA/$54.99 Canada

Covers versions 3.01 and 3.1.
Software included.

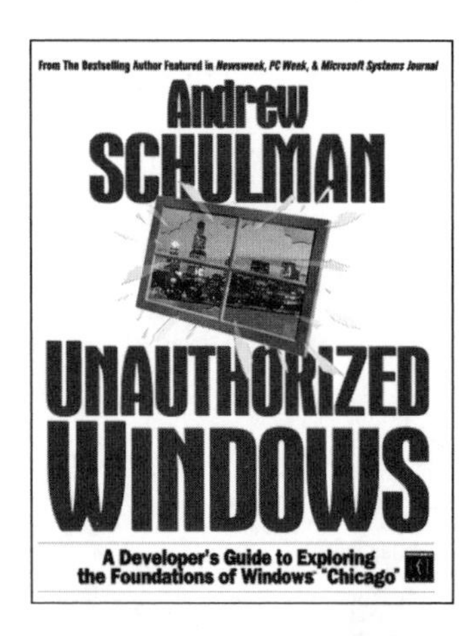

Unauthorized Windows 95: A Developer's Guide to Exploring the Foundations of Windows 95
by Andrew Schulman

ISBN: 1-56884-169-8
$29.99 USA/$39.99 Canada

Detour: The Truth About the Information Superhighway
by Michael Sullivan-Trainor

ISBN: 1-56884-307-0
$22.99 USA/$32.99 Canada

Unauthorized Windows 95: Developer's Resource Kit
by Andrew Schulman

ISBN: 1-56884-305-4
$39.99 USA/$54.99 Canada

Includes Software.

FOR MORE INFORMATION OR TO ORDER, PLEASE CALL ▸ 800 . 762 . 2974

For volume discounts & special orders please call Tony Real, Special Sales, at 415. 655. 3048

Order Center: **(800) 762-2974** *(8 a.m.–6 p.m., EST, weekdays)*

Quantity	ISBN	Title	Price	Total

Shipping & Handling Charges

	Description	First book	Each additional book	Total
Domestic	Normal	$4.50	$1.50	$
	Two Day Air	$8.50	$2.50	$
	Overnight	$18.00	$3.00	$
International	Surface	$8.00	$8.00	$
	Airmail	$16.00	$16.00	$
	DHL Air	$17.00	$17.00	$

*For large quantities call for shipping & handling charges.
**Prices are subject to change without notice.

Ship to:

Name ______________________________

Company ______________________________

Address ______________________________

City/State/Zip ______________________________

Daytime Phone ______________________________

Payment: ☐ Check to IDG Books (US Funds Only)

☐ VISA ☐ MasterCard ☐ American Express

Card # ______________________ Expires __________

Signature ______________________________

Subtotal __________

CA residents add applicable sales tax __________

IN, MA, and MD residents add 5% sales tax __________

IL residents add 6.25% sales tax __________

RI residents add 7% sales tax __________

TX residents add 8.25% sales tax __________

Shipping __________

Total __________

Please send this order form to:

IDG Books Worldwide
7260 Shadeland Station, Suite 100
Indianapolis, IN 46256

Allow up to 3 weeks for delivery.
Thank you!

IDG BOOKS WORLDWIDE REGISTRATION CARD

RETURN THIS REGISTRATION CARD FOR FREE CATALOG

Title of this book: ISDN FOR DUMMIES

My overall rating of this book: ❑ Very good [1] ❑ Good [2] ❑ Satisfactory [3] ❑ Fair [4] ❑ Poor [5]

How I first heard about this book:

❑ Found in bookstore; name: [6] ❑ Book review: [7]

❑ Advertisement: [8] ❑ Catalog: [9]

❑ Word of mouth; heard about book from friend, co-worker, etc.: [10] ❑ Other: [11]

What I liked most about this book:

What I would change, add, delete, etc., in future editions of this book:

Other comments:

Number of computer books I purchase in a year: ❑ 1 [12] ❑ 2-5 [13] ❑ 6-10 [14] ❑ More than 10 [15]

I would characterize my computer skills as: ❑ Beginner [16] ❑ Intermediate [17] ❑ Advanced [18] ❑ Professional [19]

I use ❑ DOS [20] ❑ Windows [21] ❑ OS/2 [22] ❑ Unix [23] ❑ Macintosh [24] ❑ Other: [25] ____________ (please specify)

I would be interested in new books on the following subjects:
(please check all that apply, and use the spaces provided to identify specific software)

❑ Word processing: [26] ❑ Spreadsheets: [27]

❑ Data bases: [28] ❑ Desktop publishing: [29]

❑ File Utilities: [30] ❑ Money management: [31]

❑ Networking: [32] ❑ Programming languages: [33]

❑ Other: [34]

I use a PC at (please check all that apply): ❑ home [35] ❑ work [36] ❑ school [37] ❑ other: [38] ____________

The disks I prefer to use are ❑ 5.25 [39] ❑ 3.5 [40] ❑ other: [41] ____________

I have a CD ROM: ❑ yes [42] ❑ no [43]

I plan to buy or upgrade computer hardware this year: ❑ yes [44] ❑ no [45]

I plan to buy or upgrade computer software this year: ❑ yes [46] ❑ no [47]

Name: Business title: [48] Type of Business: [49]

Address (❑ home [50] ❑ work [51]/Company name:)

Street/Suite#

City [52]/State [53]/Zipcode [54]: Country [55]

❑ **I liked this book!** You may quote me by name in future IDG Books Worldwide promotional materials.

My daytime phone number is ____________

IDG BOOKS

THE WORLD OF COMPUTER KNOWLEDGE